The Urban Ocean

This book introduces the new discipline of urban oceanography, providing a deeper understanding of the physics of the coastal ocean in an urban setting. The authors explore how the coastal ocean affects the humans who live, work and play along its shores; and in turn how human activities impact the health and dynamics of the coastal ocean. Fundamental topics covered include the governing dynamical equations, tidal and circulation processes, variation of salinity and freshwater fluxes, watershed pollutants, observing systems, and climate change. Bridging the gaps between the fields of engineering, physical and social sciences, economics and policy, this book is for anyone who wishes to learn about the physics, chemistry and biology of coastal waters. It will support an introductory course on urban oceanography at the advanced undergraduate and graduate level and will also prove invaluable as a reference text for researchers, professionals, coastal urban planners and environmental engineers.

ALAN F. BLUMBERG is an urban oceanographer who studies the interaction between cities and their offshore coastal waters. He is co-founder of Jupiter, a Silicon Valley startup company that is deeply committed to the practical application of the world's best hydroscience, weather prediction and climate modeling. The Princeton Ocean Model, developed by Blumberg and George Mellor, is utilized by scientists and institutions throughout the world. For 15 years, he was George Meade Bond Professor of Ocean Engineering and Director of the Davidson Laboratory at Stevens Institute of Technology where he led several major studies to predict and assess storm flooding events. He is the recipient of the 2001 Karl Emil Hilgard Hydraulic Prize from the American Society of Civil Engineers and the 2007 Denny Medal from the Institute of Marine Engineering. A fellow of both the American Meteorological Society and the American Society of Civil Engineers, Dr. Blumberg is the author of more than 150 journal articles. Because of his extensive research expertise, he is highly sought after by the media during unusual weather conditions.

MICHAEL S. BRUNO is Vice Chancellor for Research and Professor of Ocean Engineering at the University of Hawai'i at Mānoa and Visiting Professor in Mechanical Engineering at University College London. He is the author of more than 100 technical publications regarding coastal dynamics, ocean observation systems and community resilience. He has served on numerous advisory committees, including chairing the Marine Board of the National Academies. A Fulbright Scholar, Dr. Bruno is also a fellow of the American Society of Civil Engineers. He received the Denny Medal from the Institute of Marine Engineering in 2007, the Young Investigator Award from the Office of Naval Research in 1991 and the Outstanding Service Award from the American Society of Civil Engineers in 1988.

"Extremes are becoming more extreme in the most extreme places of our planet where too much water hits us hardest: on our coasts and along our rivers. These coastal areas increasingly urbanize, becoming more and more vulnerable to disasters, with stronger storms and rising sea levels. Blumberg and Bruno argue from their inspiring perspective of hope and belief in impactful human action that these urban hotspots along our coasts and rivers are our best opportunity for a resilient future. Here we can turn climate risks into real urban rewards. But this is only if – by design – we dare to face and better understand our climate challenges, value and manage the urban opportunities, and be radically inclusive in our approaches to produce the best solutions and increase the resilience of our urbanizing coastal regions. We'd better start now with putting their words into practice!"

Henk W. J. Ovink, Special Envoy for International Water Affairs, the Netherlands

"*The Urban Ocean* by Blumberg and Bruno is simply amazing; it is a very large compendium of facts, problem descriptions and deductions concerning the near-shore ocean and inland waters close to where many of us live. Skillfully written, it should be a valuable and unique source for scientists, urban planners, environmental managers and the curious. The use of equations where appropriate will be helpful to some readers, but will not be intimidating to those less mathematically inclined."

George Mellor, Princeton University

"Over land, sea and air, we now live in an urban world. Our cities have become ecosystems of their own, and our deltas have changed colors, transforming from a natural green to an urban red. Our oceans – for centuries a trusted resource in our daily lives – have been impacted by urbanization as well. As we strive for a more sustainable future, oceans need and deserve our attention and respect to safeguard their viability for years to come. If not, our way of life will be threatened with consequences never before experienced or imagined. I commend the authors of this fantastic and unique book that helps show us how to respect the ocean, and better understand how we ensure a resilient and sustainable future."

Piet Dircke, Global Leader Water Management, ARCADIS

"This is a unique and daring book on a fascinating and important topic. The rivers, estuaries and coastal regions of our world have lots to offer. They have therefore become increasingly urbanized. Humans have become a geophysical and a geopolitical force. The concept of *The Urban Ocean* explores and explains the formulae that govern the physics of the ocean and brings people and their habitats fully into the equation. The book bravely links physics and engineering to social studies and behavioral science. Written from an action perspective, it pairs the complex dynamics of our contemporary urbanized deltas with a globally emerging notion and movement of resilience engineering."

Theo Toonen, University of Twente

The Urban Ocean

The Interaction of Cities with Water

ALAN F. BLUMBERG
Jupiter

MICHAEL S. BRUNO
University of Hawai‘i at Mānoa

CAMBRIDGE
UNIVERSITY PRESS

University Printing House, Cambridge CB2 8BS, United Kingdom

One Liberty Plaza, 20th Floor, New York, NY 10006, USA

477 Williamstown Road, Port Melbourne, VIC 3207, Australia

314–321, 3rd Floor, Plot 3, Splendor Forum, Jasola District Centre, New Delhi – 110025, India

79 Anson Road, #06–04/06, Singapore 079906

Cambridge University Press is part of the University of Cambridge.

It furthers the University's mission by disseminating knowledge in the pursuit of education, learning, and research at the highest international levels of excellence.

www.cambridge.org
Information on this title: www.cambridge.org/9781107191990
DOI: 10.1017/9781108123839

© Cambridge University Press 2018

First published 2018

Printed in the United Kingdom by TJ International Ltd: Padstow Cornwall

A catalogue record for this publication is available from the British Library.

Library of Congress Cataloging-in-Publication Data
Names: Blumberg, Alan F., author. | Bruno, Michael S., author.
Title: The urban ocean : the interaction of cities with water / Alan F. Blumberg and
    Michael S. Bruno.
Description: Cambridge, United Kingdom ; New York, NY : Cambridge University Press, 2018. |
    Includes bibliographical references and index.
Identifiers: LCCN 2018021295| ISBN 9781107191990 (hardback) | ISBN 9781316642207
    (paperback)
Subjects: LCSH: Coasts–Environmental aspects. | Coastal zone management. | Cities and towns–
    Environmental aspects. | Coast changes.
Classification: LCC GB451.2 .B63 2018 | DDC 333.91/7091732–dc23
LC record available at https://lccn.loc.gov/2018021295

ISBN 978-1-107-19199-0 Hardback
ISBN 978-1-316-64220-7 Paperback

Alan F. Blumberg

To my parents Lily and Zelig Blumberg ז״ל and my wife Robin Blumberg ז״ל who inspired me to reach higher.

The book would not have been written without the encouragement and loving support of my children, Nathan and Jessica, and my partner, Sara Cass.

Michael S. Bruno

To all who have inspired and guided me along a richly rewarding journey, most especially my parents, Annie Marie and Frank; my brothers and sisters, Kevin, Donald, Denise, and Nancy Anne; and my wife, Cristina. You have taught me so much.

Contents

Color plate section to be found between pages 146 and 147

Preface

This book seems to us to be the inevitable product of a number of influences and events, some long in the making and others much more recent. Both of our work has for decades been focused on developing a better understanding of the dynamics of estuarine and coastal ocean regions, and the application of that understanding to the solution of problems as varied as safe navigation, water pollution, shoreline erosion and coastal flooding. Over these decades, we have witnessed profound changes in the nature and extent of these problems, and the coastal ocean environment in which we have worked. And we have seen technological advances ranging from satellite and remote sensing to high-speed computing, machine learning and artificial intelligence change the manner in which we study the ocean, and alter our understanding of the various ways in which human activities have impacted the oceans and coastal land margins around the world.

The earth has become increasingly urban in character, with the number of megacities (population greater than 10 million) quadrupling to nearly 40 over the last 30 years. A majority of these cities are located on the coastal ocean, meaning that a growing share of the world's population is not only benefiting from the climate and resources of the coastal ocean but also contributing to the myriad environmental stressors to this vital ecosystem. This also means that a growing share of the world's population lives in harm's way due to the risks associated with coastal storms and flooding, and in some areas, tsunamis. These risks have been significantly magnified by climate change, which has produced a gradual – and perhaps accelerating – rise in sea level, thereby increasing the vulnerability of coastal urban areas to catastrophic flooding and inundation. The loss of natural protective features, in particular coral reefs as a result of ocean acidification associated with carbon dioxide emissions, has further increased the risks to many coastal communities. The confluence of these two global phenomena – climate change and population migration to urban areas – as well as our passion for developing a better understanding of, and solutions to, the threats to coastal communities were the primary driving forces behind this book.

Optimism is also a driving force. The two of us have witnessed in a very personal way both the capacity for humankind to do damage to the coastal ocean environment and the capacity to mitigate or even reverse this damage. We both recall working in the Hudson–Raritan Estuary in the 1980s when concerns about water quality dominated our planning for field experiments and virtually prohibited any contact recreation in the Hudson River and its tributaries. What we remember most about this period is the profound and lasting impact that human-caused degradation of the waters and shorelines in the New York metropolitan region had on the residents' connection to the water. For nearly 100 years, the waterfront in the region was a place to be avoided; a place where commerce was conducted, but where the populace was largely absent because of health concerns. Today, that same waterfront is among the most vibrant and valuable land on earth, with parks, boat marinas, and even kayak and windsurfing launching facilities lining the Hudson River and its tributaries. Remarkably, the millions of citizens in the region have readopted the river and its tributaries as a place of beauty, a place of rest and recreation, and a place to be protected and sustained. This same story has been repeated across the planet, from Hong Kong to Vancouver to Boston; waterfronts in effect have been re-purposed as attractive gathering spaces, common areas where people can work and play in concert with the water rather than in conflict with the water. It is our strong belief that this urban waterfront revitalization will further accelerate both the resolve of coastal communities to restore and preserve their coastal environments, and the efforts of policy makers and planners to address in a pro-active way the threats posed by human activities, from local environmental degradation to global climate change.

This book began as sets of lecture notes for courses we taught in Oceanography, Introduction to Meteorology, Wave Dynamics, Coastal Engineering, Coastal Ocean Dynamics and Introduction to Estuaries at Stevens Institute of Technology where we were both faculty members for more than 15 years. The book is mathematical in the sense that partial derivatives are used to communicate the governing dynamics. But we have attempted to keep the text free of lengthy derivations, wishing to make it more about the urban ocean and not a fundamental ocean dynamics text. Our course based on this book has been taught to juniors and seniors majoring in engineering and physics, as well as first-year graduate students. For some students, Chapters 4, 5 and 7 may be challenging, and instructors may want to expand on those chapters.

People who know us also know we have an interest in numerical ocean modeling, a discipline similar to atmospheric modeling. Although the book is not about modeling, readers will quickly discover that it is in the back of our minds since the equations of motion and their boundary conditions are presented in Chapters 4 and 5 as if one were poised to set up a numerical model. Chapter 7 explores situations where certain terms in the equations of motion can be neglected, leading to analyses of currents that are very commonly observed in the offshore waters of the urban ocean. These closed-form solutions are quite useful for model validations as well.

This book, then, is intended to provide the reader with an understanding of the dynamics of the coastal ocean and atmosphere, but in the context of the people who live, work and play along its shorelines and in its waters. This context informed our decision to include topics not normally found in a text such as this, including the treatment of issues related to sustainability and community resilience to extreme events such as coastal storms and flooding, and more slowly developing threats such as climate change, water quality degradation and habitat loss. Urban oceanography, as we have termed it, is structured to be attractive to readers ranging from undergraduate and graduate students at engineering and design schools to professionals working in coastal resilience. It is also intended for anyone who wishes to discover the unique characteristics of these vital regions, the opportunities and challenges associated with the ecosystem services that these regions provide to human populations, and the technical, social and policy tools and solution paths that can be pursued to protect and preserve these waters and the populations that rely on them.

Acknowledgments

This book is a product of a lifetime of discovery, a voyage that has benefited from the help and encouragement of friends and colleagues far too numerous to list in full here. As a start, the authors wish to thank their long time mentors George Mellor and Ole Madsen.

We owe an enormous debt to our many colleagues whose ideas and insights fill this book. Collaborators like Nickitas Georgas, Tom Herrington, Richard Hires, Scott Glenn, Hans Graber, Julie Pullen, Claire Weisz, Toni Jordi, and Jim Fitzpatrick provided stimulating conversations and invaluable advice and assistance in fieldwork, data analysis, and numerical modeling over the years. Firas Saleh and Sílvia Anglès made this book a reality by helping with the figures, equation editing, and indexing. Hurricane Sandy opened our minds to a new way of looking at the coastal landscape in which we had always lived and worked, and brought us into contact with innovative urban planners, policy makers, and architects.

This book is in large part based on the lecture notes of classes we taught for several years; we thank the many students in our classes for their patient listening and their important feedback. And finally, we want to acknowledge Stevens Institute of Technology and the University of Hawaiʻi for their support of our work while writing this book.

1

Overview

People and Water

1.1   Introduction

This book provides a view of the coastal ocean from a rather unique perspective – the perspective of the urban coastal zone as the primary home of the human species. In the chapters that follow, we will discuss the physical properties of this complex domain, the dynamics that influence its movement, and its interaction with the deep ocean on one side, and landforms and tributaries on the other. We will seek a deeper understanding of the fundamental relationship between ocean and atmosphere, and thereby weather and climate, short-lived extreme events and long-term persistent phenomena in the waters that line our coasts and the skies overhead. We will discuss the range of transport and mixing processes in estuaries, shallow coastal waters, and the adjacent continental shelf, setting the stage for an examination of such dominant flow features as upwelling and downwelling, river plumes, inertial currents, and wind- and wave-driven currents. We will investigate the tide and other water level phenomena, including storm surges and tsunamis. We will address coastal sediment transport and shoreline evolution. Indeed, we will examine the full range of the physical and biogeochemical characteristics of the coastal ocean, but we will do so with the underlying aim of better understanding the relationship between the populated coastal ocean – the urban ocean – and people.

Since the dawn of civilization, the coastal ocean and the human populations that reside on and along its borders have had a complex relationship. The ocean and its tributaries serve as sources of food and energy; they provide access to waterborne transportation, trade and prosperity. They are often locations of moderate climate, enabling agriculture, recreation, and a high quality of life. But these advantages come with threats to human populations in the form of natural hazards, including tsunamis, and extreme storm events, such as tropical and extratropical storms, with their attendant storm surges and coastal flooding. For the ocean, the presence of dense populations along the coast has its negative consequences as well. Waterways, shorelines and bottom sediments are exposed to pollutants arising from human activity. These include industry, municipal wastewater, stormwater runoff, agriculture (chiefly fertilizers), spills, vessels, septic systems and other sources. Contaminants can enter the food chain and can render the water unfit for drinking and unsuitable even for contact recreation. Other human influences can have profound impacts on the very nature and configuration of the coastline. The introduction of coastal structures, the elimination of natural protective features (e.g., wetlands), and the interference with natural sand supply on both the water side (e.g., via navigation channel dredging) and the land side (e.g., river dams) can permanently alter coastal landforms, resulting in the loss of coastal islands, the landward migration of shorelines, and the loss of unique and irreplaceable habitat.

In recent decades, the advantages of living on and near the coast have combined with a global migration to urban areas to create a fundamental change to population distribution and to human vulnerability to natural disasters. Consider that in 1990, there were 10 "megacities," here defined as cities having 10 million inhabitants or more. In total, these cities were home to 153 million people or slightly less than 7% of the global urban population at that time. In 2014, there were twenty-eight megacities, home to 453 million people or about 12% of the world's urban population (United Nations, 2015). By 2017, the urban population represented more than half of the world's total population, and the number of megacities had increased to 37, representing 15% of the world's urban population (Cox, 2017). More significantly for our purposes, and not surprisingly given the benefits alluded to earlier, twenty-four of these thirty-seven megacities lie in the coastal zone (defined later in this chapter), as illustrated in Table 1.1.

The risks to coastal populations have increased dramatically by perhaps the most impactful of all feedback cycles in the ocean–human relationship: climate change. Professor E. O. Wilson, preeminent entomologist and two-time Pulitzer Prize Winner, has said "humans have become the first species who are a geophysical force" (lecture delivered at Stevens Institute of Technology, May, 2007). As the atmosphere and ocean have undergone gradual warming, the absolute level of the sea surface has risen, as a consequence primarily of the thermal expansion of seawater and the introduction of water via the melting of landbased ice cover. This sea level rise has put many populated coastal regions at risk of flooding during even minor storm events. It also threatens precious coastal groundwater sources via saltwater intrusion. The warmer waters and atmosphere appear to also be causing an increase in extreme storm frequency and

Table 1.1 *The most populous cities in the world, with coastal cities shaded in gray*

Rank	City	Nation	Region	Population
1	Tokyo-Yokohama	Japan	Asia	37,900,000
2	Jakarta	Indonesia	Asia	31,760,000
3	Delhi	India	Asia	26,495,000
4	Manila	Philippines	Asia	24,245,000
5	Seoul-Incheon	South Korea	Asia	24,105,000
6	Karachi	Pakistan	Asia	23,545,000
7	Shanghai	China	Asia	23,390,000
8	Mumbai	India	Asia	22,885,000
9	New York City Area	United States	North America	21,445,000
10	Sao Paulo	Brazil	South America	20,850,000
11	Beijing	China	Asia	20,415,000
12	Mexico City	Mexico	North America	20,400,000
13	Guangzhou-Foshan	China	Asia	19,075,000
14	Osaka-Kobe-Kyoto	Japan	Asia	17,075,000
15	Dhaka	Bangladesh	Asia	16,820,000
16	Moscow	Russia	Europe	16,710,000
17	Cairo	Egypt	Africa	16,225,000
18	Bangkok	Thailand	Asia	15,645,000
19	Los Angeles-Riverside	United States	North America	15,500,000
20	Buenos Aires	Argentina	South America	15,355,000
21	Kolkata	India	Asia	14,950,000
22	Tehran	Iran	Asia	13,805,000
23	Istanbul	Turkey	Europe	13,755,000
24	Lagos	Nigeria	Africa	13,360,000
25	Tianjin	China	Asia	13,245,000
26	Shenzhen	China	Asia	12,775,000
27	Rio de Janeiro	Brazil	South America	11,900,000
28	Kinshasa	Congo	Africa	11,855,000
29	Lima	Peru	South America	11,150,000
30	Chengdu	China	Asia	11,050,000
31	Paris	France	Europe	10,950,000
32	Lahore	Pakistan	Asia	10,665,000
33	Bangalore	India	Asia	10,535,000
34	London	United Kingdom	Europe	10,470,000
35	Ho Chi Minh City	Vietnam	Asia	10,380,000
36	Chennai	India	Asia	10,265,000
37	Nagoya	Japan	Asia	10,070,000

From Cox (2017).

intensity (IPCC, 2013). The increased concentration of carbon dioxide in the atmosphere, a primary contributor to this warming, is also causing a reduction in ocean pH ("ocean acidification"). This is in turn causing profound changes in carbonate chemistry, affecting the formation of calcium carbonate (shells) by marine organisms. Clearly there is a need to acknowledge and better understand the profound ways in which human societies have been shaped by the coastal ocean and in which the coastal ocean has in turn been impacted by the presence of humans, what we will here refer to as the "Influence Cycle."

1.2   The Urban Ocean: A Definition

We have chosen to focus our attention on those areas of the world's oceans that are located along populated coastlines. This choice has the benefit of simplifying the task of examining the physical and biogeochemical properties of these domains as compared to a treatment of the 70%+ of the planet's surface that is covered by ocean. But it certainly does not mean that our region of interest is small or that its characteristics are lacking in complexity.

We will here define the coastal zone as the interface between ocean and land, extending seaward to approximately the middle of the continental shelf (a gently sloping transition from the coastline to a depth of 100–200 m, followed in most cases by an abrupt drop in water depth at the shelf break), and inland to include all areas strongly influenced by the proximity to the ocean. This region includes benthic habitats (e.g., coral reefs), intertidal habitats (e.g., beaches, wetlands, mangroves), and semienclosed bodies of water (e.g., estuaries, bays). Simply put, we are dealing with the region extending from the landward limit of tidal influence to the middle of the continental shelf. As mentioned earlier, this is not a small region of interest. On the ocean side, continental shelves often extend to more than 100 km offshore and together represent approximately 8% of the oceans' surface area (see Figure 1.1). The coastal zone is a highly dynamic region that defines the boundary between land and ocean. It is a region rich in unique and varied ecosystems, and immensely complex in its interactions of terrestrial-, water- and atmosphere-borne energy and substances.

At what point does a coastal zone come to be defined as an "urban ocean"? We have already mentioned the fact that a majority of the world's megacities are located along coastal ocean regions. But surely these are not the only locations where humans exert significant influence over the ocean and vice versa. As illustrated in our graphical depiction of the Influence Cycle in Figure 1.2, these influences are wide-ranging.

Indeed, relatively small cities have experienced this same complex relationship with the sea, many for hundreds if not thousands of years (e.g., Rome,

Figure 1.1 Map of the world showing major topographic features. The continental shelves are indicated with the color cyan. Note the very wide continental shelves along the coastlines of Argentina, northeast North America, Southeast Asia, northern Australia, and the Arctic (Amante and Eakins, 2009). (A black-and-white version of this figure appears in some formats. For the color version, please refer to the plate section.)

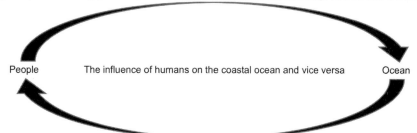

Figure 1.2 The "Influence Cycle" between humans and the coastal ocean.

Alexandria, Rotterdam, Hong Kong, Dar Es Salaam, Miami, San Juan, and Cartagena, to name a few). These regions must be included in our domain of interest. So too must regions that are not necessarily "urban" in their population density but are nonetheless home to a sufficiently large community that *the coastal ocean is markedly influenced by the human presence and is therefore decidedly different in its physical and biogeochemical properties compared to unpopulated coastal ocean regions.* This, then is the urban ocean.

Table 1.2 *Populations exposed to coastal flooding in 2005 and (predicted) in 2070*

Rank in 2070	Urban area	Nation	Exposed population in 2005	Exposed population in 2070
1	Kolkata	India	1,929,000	14,014,000
2	Mumbai	India	2,787,000	11,418,000
3	Dhaka	Bangladesh	844,000	11,135,000
4	Guangzhou	China	2,718,000	10,333,000
5	Ho Chi Minh City	Vietnam	1,931,000	9,216,000
6	Shanghai	China	2,353,000	5,451,000
7	Bangkok	Thailand	907,000	5,138,000
8	Rangoon	Myanmar	510,000	4,965,000
9	Miami	United States	2,003,000	4,795,000
10	Hai Phong	Vietnam	794,000	4,711,000

From Nicholls et al. (2008).

1.3   The Urban Ocean: The Risks

As we have already discussed, coastal communities face a number of natural hazards, including tsunamis and extreme storm events such as tropical and extratropical storms, with their attendant storm surges and coastal flooding. Flooding is in fact the most significant hazard, in terms of both loss of life and property damage (see, e.g., Doocy et al., 2013). We might reasonably expect, therefore, that urban coastal regions are high-risk areas in which to live, work, and play. But are they?

The risk equation can be stated simply as Risk = probability × consequence. We can say with some surety that the consequence of a coastal flood in a densely populated community is higher than it would be in a sparsely populated region. As an illustration, Table 1.2 lists the top ten coastal urban areas in terms of the number of people exposed to flooding, ranked by the 2070 predicted values (Nicholls et al., 2008). Table 1.3 provides a similar ranking, but according to the value of assets (buildings, infrastructure, etc.) exposed to flooding, again ranked as predicted for the year 2070. Actual values for the year 2005 are also shown in both tables. Note that six cities appear on both lists. Note also that several cities are not on our list of coastal megacities.

Because of extreme weather events along the urban coasts of the world, the loss of lives and economic damage have dramatically increased over the last decade. Four recent events are provided here as examples.

Hurricane Sandy hit the New York metro–region on October 29, 2012. Its size and direction of travel resulted in a significant storm surge, more than 3 m

Table 1.3 *Assets exposed to coastal flooding in 2005 and (predicted) in 2070*

Rank in 2070	Urban area	Nation	Exposed assets in 2005 (US$ billions)	Exposed assets in 2070 (US$ billions)
1	Miami	United States	416.29	3,513.04
2	Guangzhou	China	84.17	3,357.72
3	New York–Newark	United States	320.20	2,147.35
4	Kolkata	India	31.99	1,961.44
5	Shanghai	China	72.86	1,771.17
6	Mumbai	India	46.20	1,598.05
7	Tianjin	China	29.62	1,231.48
8	Tokyo	Japan	174.29	1,207.07
9	Hong Kong	China	35.94	1,163.89
10	Bangkok	Thailand	38.72	1,117.54

From Nicholls et al. (2008).

in some areas, along the coast of New Jersey and inside New York Harbor. Throughout the United States, more than 650,000 homes were destroyed or seriously damaged, and more than 9 million customers lost electricity. Total direct economic losses due to the hurricane have been estimated at US$72 billion (Aon Benfield, 2013). The preparation and response to Hurricane Sandy varied widely across businesses and governments. In Jersey City, for example, about 75% of the population lost power, with many residents not having gas and electricity restored for more than a week; 2,500 residents sought shelter due to lack of power, water, and heat. With 50,000 people living in one square mile, Hoboken is the fourth most densely populated municipality in the United States. Many of its residents were without power for nearly two weeks after the storm. The hurricane crippled the Port Authority Trans-Hudson line (PATH), a 24-hour subway, which in 2016 ferried 76.6 million passengers between Manhattan and New Jersey. The entire system was out for two weeks, the line to the World Trade Center was out for four weeks, and the Hoboken line was out for nearly two months. All repairs and projected costs to the PATH system are expected to ultimately exceed US$700 million.

Cyclone Phailin made landfall in October 2013 along the eastern coastline of India, in Orissa and Andhra Pradesh states. The storm affected more than 13 million people, damaged more than 300,000 homes, and caused forty-four deaths. The impacts from this potentially devastating storm were significantly reduced because of the preparation and response by emergency management authorities, including early warning, the implementation of evacuation plans (among the largest evacuations in the nation's history), the provision of cyclone shelters, and the training of thousands of first responders. This is noteworthy because in 1999, a cyclone killed more than 10,000 people in Orissa state.

On March 11, 2011, a magnitude-9 earthquake shook northeastern Japan, near Sendai. The effects of the great earthquake were felt around the world. Less than an hour after the earthquake, the first of many tsunami waves hit Japan's coastline. The tsunami waves traveled inland as far as 6 miles and flooded an area of approximately 217 square miles in Japan. The waves overtopped and destroyed protective seawalls at several locations. The Fukushima Daiichi Nuclear Power Plant suffered a level 7 nuclear meltdown after the tsunami. The total damages from the earthquake and tsunami are estimated at more than US$220 billion dollars.

Over the last decade, England and Wales have experienced frequent water-related challenges. The floods of Gloucestershire in 2007 and Somerset in 2014 in particular demonstrated the multifaceted challenges of ensuring resilience to flooding. The impacts were both local (extensive property damage and three deaths) and more widespread, with national food and transport systems severely impacted. The 2014 event caused major disruptions to road and rail systems, including the severing of the only rail line to the South West of England resulting in rail services to the west being suspended for two months. Gatwick Airport suffered severe disruption on the 23rd and 24th of December, with partial closure of its North Terminal because basement flooding knocked out key power and IT systems.

These events are painful reminders that coastal flooding is among the world's most costly and deadly disasters, capable of causing tens to hundreds of billions of dollars in damage and destroying entire neighborhoods and critical infrastructure. Coastal urban cities must become far better able to withstand, quickly recover from, and adapt to extreme weather events and flooding. And the potential for even worse flooding is on the horizon, as sea level rise causes even previously unnoticed high tide events to produce noteworthy damaging floods.

We note that simply because an urban coastal region possesses a large population and/or a high number of assets exposed to flooding, and hence would experience a high consequence as a result of a flooding event, that region's risk is not clearly understood unless we understand the probability of that flooding event. We must turn therefore to an assessment of the probability of occurrence of extreme events such as tropical and extratropical storms, and tsunamis. In March, 2015, The UN Office for Disaster Risk Reduction (UNISDR) began to implement an ISO standard for resilient and sustainable cities, ISO 37120. The UN is promoting resilience through the Sendai Framework, which was in part a global response to the 2011 Sendai earthquake, tsunami and nuclear disaster.

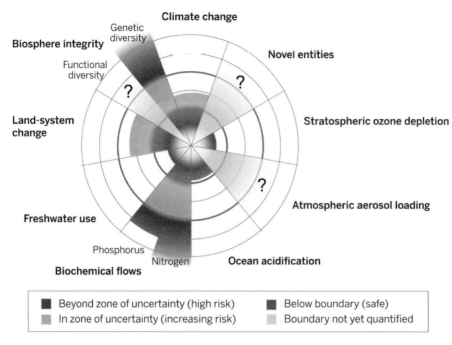

Figure 1.3 Planetary Boundaries. Reproduced from Steffen et al. (2015), with permission. (A black-and-white version of this figure appears in some formats. For the color version, please refer to the plate section.)

1.4 A Global Perspective

Rockström et al. (2009) introduced an analysis commonly referred to as the "planetary boundaries" approach. This approach aims to define the conditions under which society can thrive in a sustainable way. It can be viewed as a framework that must be updated as conditions change and as our understanding of the Earth system, with its complexities, evolves. The approach was further refined several years later (Steffen et al., 2015) and is illustrated in Figure 1.3.

The planetary boundaries can be regarded as constraints, similar to the constraints (regulations) that communities place on the amount of a particular pollutant that an industrial activity may release into the ocean. In the case of planetary boundaries, however, we are treating human influences that extend far beyond the local or regional, and instead impact the Earth system. The Boundaries define a so-called safe operating space in which humans can thrive without adversely and perhaps permanently altering certain critical Earth system processes. The assessment of the state of each process depicted in Figure 1.3 is based on an assumption that the preservation of a Holocene-like state of the Earth system is desirable. It is important to realize that impacts at local or regional scales could conceivably propagate across scales to planetary scales. The figure therefore represents a tool to guide the discussion of human influences on the Earth system, incorporating both gaps in our knowledge and uncertainty in our understanding of the functioning of the system. Nine planetary boundaries are defined. The current status of the control variables for seven of the nine planetary boundaries are shown in Figure 1.3. The green zone represents the safe operating space; the yellow zone represents the zone of uncertainty, which in order to allow for proactive response by society is termed the zone of increasing risk; and the red zone is the high-risk zone, which resides beyond the heavy circle that defines the planetary boundary. Processes for which global-level boundaries cannot yet be quantified are represented by gray wedges and question marks; these are atmospheric aerosol loading, novel entities and one aspect of biosphere integrity. With respect to biosphere integrity, two components are employed. One, E/MSY (extinctions per million species-years), employs the global extinction rate as the control variable that defines the long-term capacity of the biosphere to survive and adapt to both extreme and gradual changes in the environment. The other component, Biodiversity Intactness Index (BII), measures the loss of biodiversity by assessing changes in population abundance (across a wide range of taxa and functional groups) as a result of human impacts, using preindustrial era abundance as a reference point. Note that the treatment of biochemical flows addresses phosphorus (P) and nitrogen (N) only, although other elements as well as the ratios between elements are known to be important to biodiversity. Novel entities is a measure of the new substances and modified life-forms that can produce adverse biological and/or geophysical impacts. Research is continuing into the quantification of the boundaries for this and the other two processes. We expect many, if not all, of the control variables and their perceived limits to change as our understanding improves, and so the reader is encouraged to acquaint herself/himself with the latest developments in planetary boundaries.

It is not difficult to imagine that global phenomena are contributing to the risks and challenges facing coastal communities worldwide, even in communities that have not contributed to the disruption of the particular process, e.g., climate change or ocean acidification. The planetary boundaries assessment described here concludes that several processes, including biochemical flows, biosphere integrity, land-system change, and climate change, are beyond or near the boundary. This is particularly noteworthy for urban coastal communities. Alteration to normal or "stable" biochemical flows are often attributable in coastal and estuarine environments to excessive nutrients associated with agriculture. Impacts include so-called dead zones, or ocean regions with anoxic or near-anoxic conditions (e.g., in the Gulf of Mexico offshore of Texas), as well as harmful algal blooms. Coastal ocean habitat loss and degradation, particularly in coral reef and wetland ecosystems, and coupled in many cases with overfishing, threatens the survival of many aquatic species and hence biosphere integrity. These impacts are compounded by other processes, notably climate change and ocean acidification. The vast scale of land use change along coastal regions, including in particular deforestation, has been a major contributor to land-system change and climate change. And clearly climate change presents among the most profound threats to coastal communities, via sea level rise and the increased frequency and intensity of extreme weather events.

In future chapters, we will discuss the implications of human activity for the coastal ocean, including climate change, and the risks associated with living near and along a rapidly changing ocean. Throughout, we will take care to understand the uncertainties, both in the predictions of the severity of future coastal hazards and in the ability of human populations to respond and adapt to these hazards. We will discuss the prospects for reducing the vulnerability of urban coastal communities via science-informed public policy and urban design, enabled by improved ocean observations, understanding, and forecasting. We will introduce the concept of resilience, wherein we recognize the limits of our ability to reduce the likelihood of certain hazards and so focus our attention on learning and adapting in a way that ensures the functioning of socio-technical systems under the full range of conditions, including the expected and the unexpected.

2

Characteristics of Seawater

2.1 Introduction

In the days and weeks following Hurricane Sandy's strike on the New York City metropolitan area, people were asking why all their trees and shrubs were dying. It was an easy question to answer because of where the water that spilled onto the yards and parks of the city came from – the offshore coastal waters, which were composed of seawater. Concentrated sodium (Na), a component of sea water salt, can damage plant tissue whether it contacts above or below ground parts. High salinity can reduce plant growth and may even cause plant death. In the word "seawater," the "sea" is critical, because coastal urban areas are adjacent to estuarine and coastal waters that contain salt.

2.2 Urban Ocean Properties

The unique characteristics of the urban ocean are in many ways associated with the peculiar properties of seawater. Seawater is a mixture of various salts and water, with salinity being a measure of the total amount of dissolved salts in the water. Another important physical characteristic of seawater is its temperature. Salinity and temperature variations in the urban ocean affect the currents and water levels through their contribution to density.

Most of the water in the ocean basins is believed to originate from the condensation of water found in the early atmosphere as the Earth cooled after its formation. As the Earth's material warmed and partially melted, water locked in the minerals as hydrogen and oxygen was carried to the surface by volcanic venting activity. This water was released as volcanic outgassing from

Table 2.1 *Sources of salt in the ocean*

- Weathering rock via rivers
- Volcanoes, hot springs
- Hydrothermal vents
- Rain and atmospheric deposition
- Meteors from space

the lithosphere as the Earth's crust solidified. It is also thought that some of the Earth's water came from comets from outer space.

The properties of water that matter the most for oceanography are salt and water temperature, with pressure playing a minor role. Each of these factors will be discussed in turn.

The main source of salt in the ocean is the erosion of rocks due to rainfall runoff. Further sources include the eruption of volcanoes, hot springs and hydrothermal vents (underwater hot springs); rain and atmospheric deposition; and meteors from space (Table 2.1).

The total concentration of dissolved salts varies significantly from place to place in the ocean. These differences result from dilution by freshwater from river runoff and rain, and from evaporation.

2.3   Salinity

In the coastal waters of the world, about 3% of the weight of seawater is dissolved salt and the other 97% is water. In contrast, the open ocean salinity is about 3.5%. A typical kilogram (1,000 g) sample of seawater is composed of 970 g of water and 30 g of salt. Salinity is difficult to define and to measure, so to avoid this difficulty, oceanographers use conductivity instead of salinity. They measure conductivity and temperature and calculate density from these variables.

Seawater is a complicated solution containing the majority of the known elements. Some of the more abundant components are listed in Table 2.2 as a percent of total mass of dissolved material.

These six salts make up more than 99% of the salts dissolved in seawater. If the water in a sample of seawater were to evaporate, these ions would combine to form a solid precipitate. For example, sodium Na^+ and chlorine Cl^- would form the salt compound sodium chloride, NaCl. It is easy to see from Table 2.2 that sodium and chlorine account for 86% of the dissolved constituents in seawater.

Table 2.2 *Major ions in seawater by percentage by weight*

Compound	Chemical symbol	Percentage by weight
Chlorine	Cl^-	55.3%
Sodium	Na^+	30.8%
Sulfate	SO_4^{++}	7.7%
Magnesium	Mg^{++}	3.7%
Calcium	Ca^+	1.2%
Potassium	K^+	1.1%

Salinity (S) was originally defined as the mass in kilograms of solid material in a kilogram of seawater after evaporating the water away. It is the mass of dissolved solids in one kilogram of seawater divided by one kilogram of seawater. Salinity is a dimensionless variable – mass divided by mass. That is why we say, officially, that salinity has no units. Salinity is expressed in parts per thousand or, in virtually synonymous terms, practical salinity units (psu). Salinity is difficult to measure gravimetrically because many of the salts are hydrophilic and some decompose on heating to dryness. From about 1900 to the 1960s, salinity was calculated from chlorinity as determined by titration with silver ions. As of 1978, it became standard to calculate "practical salinity" via electrical conductivity and make an equivalence to salinity. The conductivity is directly proportional to the content of dissolved salts and can therefore be parlayed into a salinity determination.

The ocean's average salinity is about 35 psu, but the salinity in the urban ocean can increase from 0 within the river source waters to 35 psu as we move away from the shallower waters toward the open ocean. The greater the salinity, the heavier the water, because dissolved salt adds to the mass of the water and makes the water denser than it would be otherwise. River runoff has the direct effect of reducing salinity in the surface layer in areas where mixing is not significant and throughout the water column where mixing occurs. In temperate and tropical climates, the runoff from rivers will be at a maximum during the rainy season and will decrease the salinity significantly below the ocean's average of 35 psu.

2.4   Temperature

Water temperature is a direct measure of the kinetic energy of the water molecules in a substance: the higher the temperature, the faster the molecules are moving. The measurement of temperature using an absolute scale is challenging, and so it is usually made by national standards laboratories. The absolute measurements are used to define a practical temperature scale based on the temperature of a few fixed points and interpolating devices that are calibrated at the fixed points.

We usually use degrees Celsius or centigrade for temperature rather than Kelvin or Fahrenheit. Care should be taken when converting, for example, Celsius to Fahrenheit and vice versa. A quick rule of thumb when converting temperatures from Celsius to Fahrenheit is to double (actually 9/5) the Celsius value and add 32. Conversely, when converting from Fahrenheit to Celsius, subtract 32 from the Celsius value and divide by 2 (actually 9/5 again). Joules or calories are reserved for heat content. Calorie is the amount of heat required to raise one gram of water by one degree centigrade. And we use watt to describe the rate of time change of heat: one watt is equal to one joule per second.

Temperature changes little in the urban ocean both horizontally and vertically. Temperature in the open ocean is typically well mixed in the upper 100 m or so. As this water moves into the shallow urban ocean, it brings with it a well-mixed water mass. The shallow water permits very great ranges in temperature to occur on a yearly and even a daily basis. Sea ice forms in the urban ocean, especially in estuaries in high-latitude regions.

Of all the naturally occurring Earth materials, water can hold the largest amount of heat per degree of temperature change. This property, the heat capacity, is very high compared to that of land and the atmosphere. This influences the climate of cities near the oceans, which tend to have moderated, less extreme temperatures than inland cities. This property of water is one reason why coastal urban areas and those far from the coast can differ so much in temperature patterns. Due to this high heat capacity of water, most of the added heat resulting from greenhouse gas emissions is being captured in the world's oceans, which buffers the actual rate of change of the temperature in the surface atmosphere and that of cities. A related interesting characteristic of water is its tendency for thermal expansion. As water molecules heat up they get farther apart, and the water expands to occupy more volume. For example, 1 L of water at $20°C$ will occupy 1.013 L of volume at $80°C$. As the ocean warms, the density decreases and thus even at constant mass, the volume of the ocean increases. This thermal expansion (or steric sea level rise) occurs at all seawater temperatures and has been one of the major contributors to sea level changes during the twentieth and twenty-first centuries. Water at higher pressure (deeper depths) expands more for a given heat input, so the global average expansion is affected by the distribution of heat within the ocean.

2.5   Density

Density, usually denoted by ρ, is the amount of mass per unit volume and is expressed in kilograms per cubic meter. A common way to express density involves a shorthand method called sigma. Sigma is the density minus 1,000 kg/m^3. In general terms, it is

$$\sigma(S, T) = \rho(S, T) - 1{,}000 \, \text{kg/m}^3, \tag{2.1}$$

where T is the temperature in Celsius and S is the salinity in either ppt or psu. Sigma, usually called σ_t or "sigma t," is useful because it is easier than density to write and discuss. For example, if the density of seawater is 1,027 kg/m^3, you can just say σ_t is 27 or 27 kg/m^3. It is much easier than saying 1,027 kg/m^3.

A fascinating factor differentiating seawater from pure water is that ice easily forms on the surface of pure water, whereas it is much more difficult for ice to form on seawater. This is because pure water exhibits a well-known anomalous density maximum at 4°C, whereas this does not occur in seawater of salinity greater than 24.7 ppt. When surface water in a lake is cooled, its density increases slightly until it reaches a density maximum at 3.98°C. After that temperature is reached, the density decreases as the temperature both increases and decreases. As pure water cools below 4°C, it floats, whereas seawater cooled below 4°C continues to sink. That sinking seawater is replaced by water from below that is warmer and thus lighter, making it much more difficult for ice to form on seawater, though it does occur in the polar oceans.

The presence of salt decreases the freezing point of water and decreases the temperature of maximum density. The freezing point of pure water is 0°C but the freezing point of seawater with a salinity of 35 ppt is −2.3°C. In the initial stage of ice formation on salt water, salt is rejected and increases the density of the surrounding seawater, some of which begins to sink. This is known as brine rejection. The increase in salinity can be substantial and can result in the formation of downward cascades of dense saline waters.

2.6   Equation of State

Density is a function of temperature, salinity, and pressure. Since the water depths we study in urban oceanography are less than 200 m, the pressure contribution is very small and typically neglected. As temperature increases, density decreases. As salinity increases, density increases. The relationship between temperature and salinity is complex, weakly nonlinear, and captured by an unwieldly mathematical relationship. Figure 2.1 shows the changes in σ_t with respect to temperature and salinity. The x axis is salinity, and the y axis is

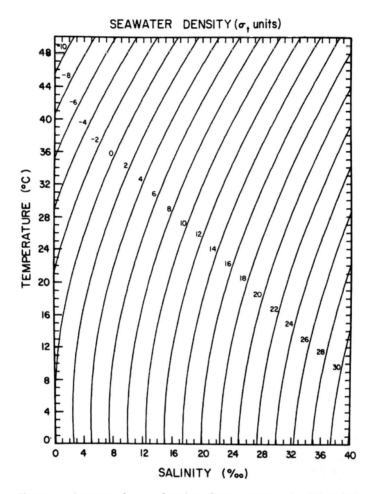

Figure 2.1 Contours of σ_t as a function of temperature and salinity (Fischer et al., 1979).

temperature. The lines on the graph are not linear but are curved, emphasizing the complexity of the relationship of temperature and salinity to density. For example, if the salinity of a seawater sample is 20 ppt and the temperature of that sample is 20°C, then by using this graph, you can easily estimate σ_t to be about 13 kg/m³. In other words, the density is about 1,013 kg/m³.

Figure 2.1 also demonstrates that salinity has a nearly 1-to-1 influence on density, whereas it requires several degrees of temperature change to create a change of 1 kg/m³ density. As the salinity increases, σ_t increases as well. And as the temperature increases, σ_t decreases. In the urban ocean, salinity dominates the density variations for the most part. A rule of thumb when considering warm water is that a change of 5°C is equivalent to a change of 1 ppt in density. In the

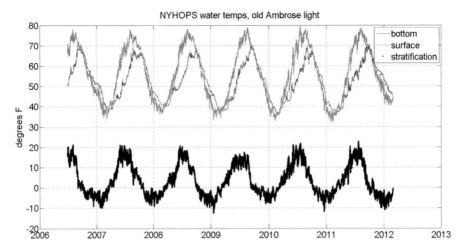

Figure 2.2 Surface and bottom temperatures and the temperature stratification. (A black-and-white version of this figure appears in some formats. For the color version, please refer to the plate section.)

urban ocean, salinity then becomes the controlling constituent as the freshwater river inputs are measurably diluting the ocean waters, leading to a wide range of salinity values.

To illustrate the complexity of the density as a function of temperature and salinity, consider the times series of surface and bottom temperatures off the coast of New Jersey over a six-year period, as shown in Figure 2.2. The difference between the surface and bottom temperature is also shown. That difference is called the stratification. The question is now, why is the surface temperature sometimes colder than the bottom temperature in winter and the bottom temperature colder than the surface temperature in the other seasons? How could this be physically true since cold water is heavier than warm water?

The density of the water controls the stability of the water column, not temperature or salinity. As a rule, the density of water is either constant or increases with depth. If the density of the surface water exceeds that of the water below it, the situation is gravitationally unstable and the surface water sinks. The density depends on both temperature and salinity. When the bottom water is warmer than the surface water suggesting an imbalance, it means the bottom water must be much saltier than the surface water to keep the density stable.

2.7   Sound in the Urban Ocean

In the atmosphere, we can see much farther than we can hear. In the ocean, the opposite is true, as humans can see objects typically no more than

50 m away from them. Sound, on the other hand, can be detected over vast distances. For typical oceanic conditions, the speed of sound is usually between 1,450 m/s and 1,550 m/s. The approximate variation is: 4 m/s per 1°C rise in temperature; and 1.5 m/s per 1 psu increase in salinity. The primary cause of variability of sound speed in the urban ocean, unlike the open ocean, is salinity. Variations of temperature are typically too small to have much influence.

2.8   Light in the Urban Ocean

Sunlight with its range of wavelengths enters the ocean after passing through the atmosphere. Within the upper layers of the water column, up to 100 m or so, the visible light interacts with the water molecules and substances that are dissolved or suspended in the water. The incoming sunlight is primarily determined by latitude, season, time of day, and cloudiness. The polar regions are illuminated less than the tropics, areas in winter are heated less than the same area in summer, areas in early morning are heated less than the same area at noon and cloudy days have less sun than sunny days. The heat gained from sunlight in the upper layers of the water column is transmitted downward by turbulent mixing from the wind and currents.

Visible light is attenuated as it passes through water. In clear ocean water, there is sufficient light to about 50–100 m to see, but in typical coastal waters almost all of the light energy is absorbed by 10 m. Blue light penetrates farther into seawater giving the ocean its distinctive color. Only 2% of blue light is absorbed in the top 1 m of the water column. At the same time, seawater absorbs red, orange, and yellow wavelengths quickly with depth, as shown in Table 2.3, removing these colors. If there are phytoplankton in the water, their chlorophyll absorbs the blue and the red light, which shifts the color of the water to green,

Table 2.3 *Penetration of light of various wavelengths into seawater*

Color	Wavelength (nm)	% absorbed in 1 m of water	Depth at which 99% is absorbed (m)
Infrared	800	82.0	3
Red	725	71.0	4
Orange	600	16.7	25
Yellow	575	8.7	51
Green	525	4.0	113
Blue	475	1.8	254
Violet	400	4.2	107
Ultraviolet	310	14.0	31

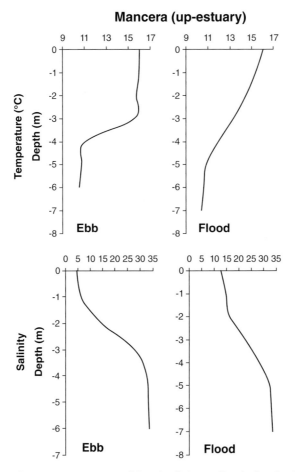

Figure 2.3 Temperature (°C) and salinity profiles during the ebb and flood tidal phases at Mancera (Ochoa-Muñoz et al., 2013).

yellow and brown, and sometimes even black appearance. Thus, most of the light that hits the surface of the water is absorbed or scattered within the top few meters of the ocean. In fact, only 73% of the surface light reaches a depth of 1 cm, only 44.5% reaches a depth of 1 m and just 22.2% reaches a depth of 10 m.

2.9   Some Vocabulary

A typical example of a vertical profile of salinity in an estuary is shown in Figure 2.3. The region of the greatest change in salinity is called the halocline, and for temperature it is called the thermocline. In Figure 2.3, both the thermocline and halocline occur about 3.5 m below the surface. The thermocline and the halocline are much less pronounced during the flood cycle when water is coming in from the sea.

The pycnocline encompasses both the halocline (salinity gradients) and the thermocline (temperature gradients). Stratification refers to the arrangement of water masses in layers according to their densities. Water density of course increases with depth, but not at a constant rate. In the urban ocean, the water column is generally characterized by three distinct layers: an upper mixed layer; the halocline/thermocline (a region in which the salinity/temperature increases/ decreases and density increases rapidly with increasing depth); and a deeper zone of well-mixed water, in which density increases slowly with depth. Stratification, it should be mentioned, forms an effective barrier for the exchange of nutrients and dissolved gases between the top and the deeper, nutrient-richer waters. Stratification therefore has important implications for biological and biogeo-chemical processes in the urban ocean. For example, in coastal waters, where the flux of settling organic matter is high, prolonged periods of stratification can lead to hypoxia (low oxygen), causing mortality of fish, crabs and other marine organisms.

3

Urban Ocean Characteristics

3.1 Introduction

The urban ocean is the place where the ocean, the land, and the people all come together. Each of these component "systems" has a profound effect on the others. The coastal ocean and associated weather dynamics drive processes and events that range from highly supportive of human populations (e.g., via fishing, marine transportation, and tourism) to highly threatening (e.g., storm surges and flooding). The tide creates complex dynamics, often in the context of an estuary environment, that influence the physical and biogeochemical characteristics of the region and thereby define its capability and capacity to support human life. This balancing act between ocean-as-sustainer and ocean-as-threat has produced a wide range of coastal ocean "management" strategies that have themselves often resulted in significant short- and long-term changes to the balance. The dredging of shipping channels intended to support safe navigation and economic prosperity has often led to the alteration of tidal and wind-driven dynamics and transport processes, with significant consequences to ecosystems both inland and along the coast. The construction of sea walls to protect life and property often causes beach erosion and long-term shoreline retreat. These are only a few of the myriad ways in which the symbiotic relationship between humans and the coastal ocean has produced significant impacts, many of which were not expected and some of which are still now poorly understood.

LOICZ (2005) presented a comprehensive list of human influences in the coastal region. These influences include the following:

- alterations to water discharges from rivers
- water extraction for urban development, industry and agriculture
- energy fluxes within systems of the coastal domain

- regional decreases in the delivery of sediment through entrapment within reservoirs
- regional increases in the delivery of sediment through increased soil erosion
- flux of nutrients
- flow patterns and sedimentation in estuaries due to dredging and reclamation
- loss of space and habitat, and increased loadings on estuaries through sedimentation
- building of shoreline engineering structures, ports and urban developments
- harvesting, often over-harvesting, of marine resources
- loss of traditional food resources and environment (cultural values) for indigenous people
- increased competition for marine space
- increased pollutants, contaminants and atmospheric emissions from industries and urbanization
- modification of the type and quantity of coastal discharges from surface and groundwater flows
- alienation of coastal wetlands and other valuable ecosystems through land use change
- modification of habitat structure and functioning through introduction of non-indigenous species

A major challenge to the development of an effective understanding of the relationship between human activities along the coast and the health and function of the coastal ecosystem is the issue of scale. Both sides of the "equation" – human populations and ecosystems – are governed by complex dynamics that include influences acting across a vast range of temporal and spatial scales. Population changes and land use changes (e.g., from agriculture to industry) clearly place stresses on the natural functioning of land and ocean ecosystems, but the impacts vary wildly in scale. These can range, for example, from the immediate, relatively small-scale impacts of a localized oil or chemical spill to the long-term (or even permanent), large-scale impacts of the removal of sand supply to the coast via the construction of dams, and subsequent shoreline recession, a phenomenon that can take years to manifest itself. Socio-economic factors drive human behavior that can significantly alter coastal ecosystems. Consider, for example, the port cities that were established around the world as a result of population demands for certain fish (e.g., Cod), leading to economic growth and in many cases the expansion of ship-based trade. The long-term consequence to the ecosystem has all too often been overfishing and the collapse of the fishery. Continued demand for fish protein has in many cases then led to the adoption of new species as preferred food, including nearshore and reef fish, or to the development of aquaculture techniques, all with attendant impacts to nearshore water quality and ecosystem functioning. Many consequences of human activities were only understood many years after the fact, e.g., the impact of fertilizer on coastal and estuary ecosystems via excessive nutrients that can lead to eutrophication, harmful algal blooms, and hypoxia (oxygen levels below that which is required to support aquatic life). Of course, as mentioned in Chapter 1, the most significant of impacts associated with human activities, and the largest in terms of scale, is climate change. Our response to these challenges will require both socio-technical and socio-economic approaches, bearing in mind the complexities and the scales involved across the global human population and the diverse range of coastal ocean ecosystems.

In this chapter, we will discuss the characteristics of the urban ocean from the standpoint of how human populations seek advantage from being located near the ocean and in so doing how they have altered the natural system, with attendant consequences for the risks of disruption. As has already been described in Chapter 1, the urban ocean is perhaps the most complex of environments. It is a region that is fundamentally governed by boundaries: the boundary between land and water (often including both land–ocean and land–estuary boundaries); the boundary between the built environment and the natural environment; and the boundaries between a striking number of ecosystems, including terrestrial, marine, freshwater, and marsh. Our approach here will involve the examination of several of these boundaries.

3.2   The Land–Ocean Boundary

The boundary between land and ocean is a region of extraordinary bioproductivity and biodiversity. This is of course related to the exchange of materials, chemicals, and organisms between the terrestrial side of the boundary and the marine side. Sediment washed out to sea during rainfall events creates and sustains coastal beaches and – over time – dominant landforms such as headlands, dunes, and offshore bars that provide natural protection to the low-lying areas immediately landward. In coral reef environments, coral-derived sediments arise from both physical re-working of coral skeletons (e.g., as a result of wave action) and bio re-working (e.g., the remnants of feeding by reef fish). In the same way as terrestrial-derived sediments, these coral reef-derived sediments, and the reefs themselves, provide natural protection to the low-lying areas immediately landward of the coast.

In an urban ocean setting, the protection afforded by these beaches and coastal landforms is critical to the sustainability and resilience of the coastal populations that reside along and landward of the land–ocean boundary. As we will learn in later chapters, there has recently been a recognition across the international community not only of the importance of these "coastal protective features", but also the fact that these natural features are in the long term much more effective than human-built protective structures and devices. There is of course the added economic and quality-of-life benefit that beaches afford to coastal urban centers.

A significant contributor to the bio-productivity and biodiversity of the land–ocean boundary is the introduction of freshwater to the marine environment. Brackish water (slightly saline water, with salinity often variable over timescales ranging from tidal to seasonal) supports diverse and productive ecosystems in environments that can include estuaries, coastal inlets, embayments, wetlands, and mangroves. These environments will be well known to most readers, of course. But the complexity of the land–ocean boundary is evidenced by yet another type of freshwater introduction to the coastal zone – freshwater ground-water flows emanating from the ocean floor. For illustrative purposes, and also to introduce the land–ocean boundary to the reader in more detail, let us take an example from one region of the world – the Hawaiian Islands.

3.2.1   A Case Study in Complexity – the Hawaiian Islands

The land–ocean boundary along the south shore of the island of Oʻahu in Hawaiʻi is an example of an urban ocean environment. It is also a region with significant complexity along its land–ocean boundary. This is the location of the city of Honolulu, a world known tourist destination and a financial and cultural center in the Pacific region. It is also a unique coastal zone from many perspectives.

The land presently occupied by Honolulu and the mountain area immediately landward has played a prominent role throughout Native Hawaiian history. This is an area with an abundance of freshwater flowing from the mountains (where rain occurs nearly daily) toward the shoreline at Waikiki Beach. In fact, the name "Waikiki" is actually Hawaiian for "leaping water", a term not meant to describe the famous surfing waves, but rather the freshwater that would spring up out of the ground along the wetland areas at the coast! The Native Hawaiians settled here because of access to freshwater, to fishing, and to navigation – reasons that are common to early coastal settlers around the world. By the late 1700s the plentiful fresh water on Oʻahu had been used to build 113 fishponds, covering 4,200 acres (Kameʻeleihiwa, 2016). Around this time, Oʻahu was at minimum producing 1.3 million pounds of fish per year, not counting the fish obtained from the reefs and beyond (Keala, 2007).

Figure 3.1 View of the coastline in the vicinity of Honolulu, on the island of O'ahu, Hawai'i (NOAA, 2018a). (A black-and-white version of this figure appears in some formats. For the color version, please refer to the plate section.)

The Native Hawaiians understood the fragile nature of the environment in which they lived, and they employed a uniquely sustainable approach to land use (there was no land ownership), agriculture, and fishing. They understood the concept of watershed management. Ahupua'a is a Hawaiian word that is used to describe a land area that begins at the mountain peak and runs all the way to the sea. The lateral boundaries of these land areas are the ridgelines of the valleys, and hence, each Ahupua'a functions as a watershed. The Native Hawaiians understood the critical importance of water, and in fact the protection of water was considered a sacred obligation. As a result, the residents of any given Ahupua'a took care to only use the water within their own watershed (Kame'eleihiwa, 2016).

The modern city of Honolulu is shown in the photo in Figure 3.1. Note the complex built environment adjacent to the land–ocean boundary. Note also the mountainous areas in close proximity to the shoreline and the coral reef areas immediately offshore.

What is *not* visible in this photo is the continuous flow of freshwater from undersea groundwater sources all along this coastline. In Hawai'i and other mountainous islands, submarine groundwater discharge (SGD) can be a significant source of freshwater and dissolved matter to coastal ocean environments including nearshore reefs. Nelson et al. (2015) conducted a study in which they traced the dissolved substances carried seaward via SGD at two locations in Oʻahu just east of the city of Honolulu and showed how the SGD emanating from the spring at each location diffuses seaward to the reef and beyond. There is some urgency to developing a better understanding of SGD in all coastal environments, in particular because of its potential as a pathway for the migration of contaminants and excessive nutrients into the marine environment (see, e.g., Zhang and Mandal, 2012). In tropical coastal regions, SGD often carries with it naturally occurring organic material, including the nutrient nitrogen. However, in populated coastal zones, human activities such as fertilizer-assisted agriculture can dramatically increase the loading of nitrogen from these discharges. Other human activities, including wastewater injection into the ground can significantly increase the number and impacts of contaminants introduced to the marine ecosystem. This impact can be most dramatic in environments such as the Hawaiian Islands, where water is plentiful and the terrain is favorable to substantial groundwater flow toward the sea (see, e.g., Amato et al., 2016). Of course, this issue is not restricted to tropical island environments. SGD has been connected to water quality issues around the globe, including harmful algal blooms, species shifts, and eutrophication (see, e.g., Lee and Kim, 2007).

Honolulu is also an interesting case study in terms of the risk of disruption to the population as a result of natural events. Hazards exist from both sides of the land–ocean boundary. The region is prone to periods of heavy rainfall, particularly as a result of tropical cyclones. High rates of rainfall combined with steep mountain slopes can produce flash floods throughout the region, including in heavily populated areas. Figure 3.2 illustrates the type of flash flood common during a major rainfall event, in this case associated with the passage of tropical storm Darby in July, 2016. Note that the discharge rate of this relatively minor stream increased by more than three orders of magnitude during the storm! The event caused major flooding of roadways across the region, as well as the erosion of stream banks and the destruction of several homes.

From the ocean side, the region is at risk for tsunamis. For example, the 2011 Tōhoku earthquake produced a damaging tsunami in the Hawaiian Islands, with waves exceeding 1 m along the shoreline of the island of Hawaiʻi. If the wave heights experienced along the Honolulu waterfront had been this magnitude, rather than the 0.5-m height actually experienced, the damage to the properties and beaches along this economically vital shorefront would have been significant.

As a consequence of these risks, and in many ways contrary to the sustainable practices of the Native Hawaiian people, there began in the mid-1800s an effort to construct various forms of coastal protective features in Honolulu. To guard against land-side flooding, streams were channelized and, in some cases,

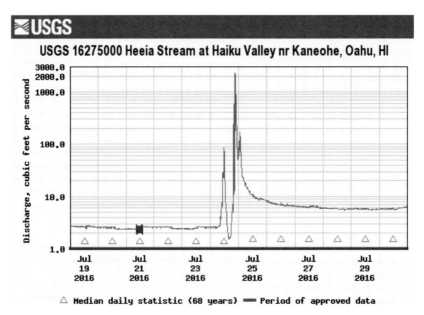

Figure 3.2 US Geological Survey (USGS) stream gauge record at Heeia Stream, Hawai'i, July 19–29, 2016 (USGS, 2018).

deepened, and the coastal wetlands were filled. To guard against ocean flooding, seawalls were built in front of property deemed economically important. Some of these structures are visible in Figure 3.1.

All of these natural and man-made influences exist in an environment with an extraordinarily rich ecosystem. The Hawaiian Islands, including the waters directly offshore of the city of Honolulu, host a diverse community of fish, turtles, and marine mammals including dolphins, various species of whales, and the Hawaiian monk seal. Because of habitat loss and the unsustainable fishing of certain fish species, there has been considerable effort made in recent years to protect the coastal ocean environment in and around the city. This includes the designation in 2007 (and in 2016 the expansion) of the Papahanaumokuakea Marine National Monument (the strongest preservation designation in US law) that will protect a 1,508,870 km^2 region immediately west of Honolulu.

3.3   The Land–Estuary Boundary

Estuaries – the places where rivers meet oceans – are among the most productive ecosystems on Earth, in large part because this is where freshwater and ocean saline water come together under the complex driving forces of river outflow, meteorological effects, and the tide. The resulting transport and mixing processes, the associated time-variable water salinity and temperature, and the space- and time-variable loadings of biogeochemical substances and biota all combine to produce a highly dynamic ecosystem characterized by a very wide array of habitats. These habitats can range from wetlands and mudflats to mangroves and sandy beaches; from bottoms composed of coarse sand to bottoms composed of very fine organic material. Not surprisingly, given their biological productivity and proximity to both river and ocean, estuaries are the primary location for human settlement along the coast, as evidenced by the fact that most of the coastal megacities listed in Table 1.1 are located along an estuary.

Scientific studies of the dynamics of estuaries include the early work of Pritchard (1954) on the salt balance in the James River Estuary, the work of Hansen and Rattray (1966) on the classification of different estuary types, Fischer's (1972) treatment of mass transport mechanisms, and the more recent work of Lerczak and Geyer (2004) on lateral circulation. The study of estuaries continues to evolve as the sophistication of observing system technology and computational analysis continues to improve.

The combination of a diverse, productive, and fragile ecosystem with a high concentration of human populations produces the risk that human activities may reduce or eliminate the ecosystem benefits (e.g., food and clean water) and protective functions (e.g., wetland buffers against wave action and storm surge) that first attracted human settlement. Activities that produce habitat alteration, the alteration of groundwater and surface water flows, interference with the transport of sediment, and the introduction of contaminants, all produce effects that threaten the viability of estuary ecosystems. As we did earlier for the land–ocean boundary, we will here introduce the reader to the land–estuary boundary by examining one example, in this case a highly urbanized estuary – The Hudson–Raritan Estuary.

3.3.1   *A Case Study in Human-Altered Estuaries – the Hudson–Raritan Estuary*

The Hudson–Raritan Estuary has been a vital waterway that has sustained significant human populations dating back to settlements along its banks by the Lenape Indians and other Native American tribes. Its use as a critical port in global commerce dates back to the Dutch East India Company in the early 1600s, after Henry Hudson navigated the estuary on behalf of the company. Today, the estuary is of enormous ecological and economic importance to the New York metropolitan region. In fact, the Port of New York and New Jersey is the largest port on the East Coast of the United States and is central to the economy of the region.

Figure 3.3 Aerial photo of the Hudson–Raritan Estuary (courtesy Stevens Institute of Technology).

The estuary has, since the early 1900s, suffered adverse impacts from industry located along its banks and the banks of its tributaries, as well as from the introduction of wastewater and debris from the densely populated communities in the region. The presence of toxic chemicals in both the water and sediments resulted in reduced water quality, fisheries restrictions/advisories, reproductive impairments in some species, and general adverse impacts on the estuarine and coastal ecosystems. As an example of the positive change that can occur when public policy, technological improvements, and socio-economic forces all come together to advance an important goal, the waters coursing through the Hudson–Raritan Estuary are today significantly cleaner than they were prior to the passage in 1972 of the Federal Clean Water Act.

The Newark Bay side of the Estuary, shown in Figure 3.3, is the primary location of the Port of New York and New Jersey. Consistently ranked among the top ports in the world in terms of tonnage, cargo value, and number of containers, the port plays an essential role in the local, national, and global economies.

The complex estuarine system in the vicinity of the Port consists of two freshwater inputs from the north, the Passaic and Hackensack Rivers, and two tidal straits. These straits are the Kill van Kull, which connects Newark Bay to the Hudson River, and the Arthur Kill, which joins Newark Bay to Raritan Bay.

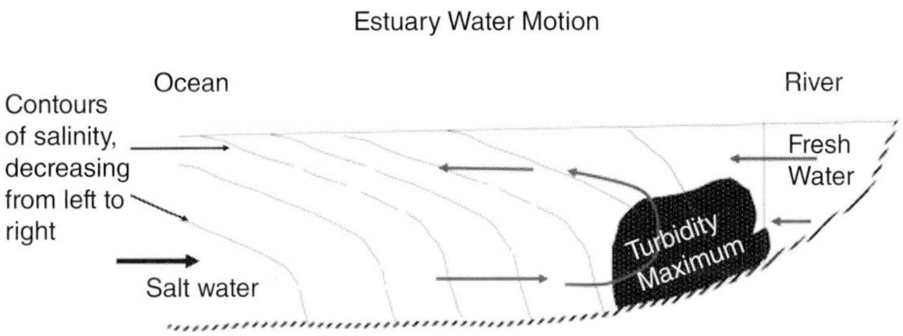

Estuary Water Motion

Figure 3.4 Illustration of the dominant flow pattern in an estuary. In estuaries, heavy salt water from the ocean flows landward underneath the lighter fresh water from the river. This results in converging near-bottom flows that accumulate suspended material near the head of the salt water flow (referred to as the salt "wedge").

The depth of this complex is naturally shallow, and shipping channels must be dredged throughout the system to maintain a depth sufficient for the safe navigation of ships arriving at the port. An agreement that was finalized in 2004 authorizes the US Army Corps of Engineers to award contracts to deepen the Kill van Kull, Newark Bay, and the Arthur Kill shipping channels to 15.2 m (50 feet) to accommodate larger container ships with deeper draft. The maintenance of the shipping channels in this area will continue to be an ongoing project, since the system tends to revert back to its naturally shallow state. Most critically, the sediments in the area are contaminated with a variety of toxic chemicals (Pecchioli et al., 2006), which leads to a high cost for the disposal of these sediments.

The results from a comprehensive observation and computer model analysis (Pence et al., 2005) showed that the hydrodynamics in the Newark Bay system are highly complex, with many external forces affecting the expected tidal flow patterns in the system. In the Newark Bay navigation channel, classical two-layer estuarine circulation is observed, wherein seaward freshwater flow from the Passaic and Hackensack Rivers is concentrated at the surface, and landward ocean water flow is concentrated at the bottom, as depicted in Figure 3.4. Note that we have included a depiction of a representative location of the "turbidity maximum" in the figure, as this is an important feature of estuarine dynamics. Turbidity is the measure of the clarity of the water. In estuaries, high turbidity (low clarity) would normally be associated with the presence of a high amount of material suspended in the water column. The phenomenon of a localized region of very high turbidity (the turbidity maximum) in most estuaries is caused by two factors. First, the physical confluence and mixing of the bottom salty water and the surface fresh water causes a high amount of turbulence, which stirs up and suspends the bottom sediments. Second, the presence of salt water causes flocculation of the very fine (organic) material in the water, resulting in larger suspended matter and an even higher level of turbidity. The location of the turbidity maximum moves upstream and downstream depending on the relative strength of the tidal flow from the ocean and the seaward river flow.

As often happens in estuaries, during periods of very low river flow, the currents in the Newark Bay channel are directed landward at all depths, as the tidal flow dominates. Interestingly, along the very shallow side banks that line the dredged channel, there is a persistent flow downstream at all times. Large, persistent wind events in the region can have a strong effect on the circulation in the estuary and in some extreme cases, can disrupt the expected pattern of estuarine flow. Strong winds from the west cause a very strong flow seaward from Newark Bay and out toward the Hudson River via the Kill van Kull, resulting in a "flushing" of the river water and its suspended material out to the Upper Bay and, depending on the tidal flow, toward the Hudson River or the ocean. This flow changes direction when strong winds are from the east. An analysis of the data shows that these flows, caused by an east or west wind, are the dominant modes in the system.

Hydrodynamic events such as the flushing events described above can have a significant effect on contaminant transport, in particular because the Passaic River contains numerous former industrial sites along its banks, many representing ongoing sources of contaminants, including dioxins and mercury. Studies have shown that the highest concentrations of these contaminants occur during periods of highest freshwater flows (Dimou et al., 2006). These analyses are critical to the understanding of the fate of contaminated sediments in the region. Large flow events from the Passaic River produce higher suspended sediment concentration in Newark Bay. The fate of this suspended sediment depends in part on the settling rate of the suspended material. High river discharge events increase the flow in the landward-flowing bottom layer in Newark Bay, which effectively traps the suspended sediment that rapidly settles to the lower layer. However, these high flow events also increase the surface outflow and can transport the slowly settling sediment toward the Kill van Kull where stronger tidal currents can easily carry this material into the Upper Bay. The details of the partitioning of contaminants across sediment size and settling velocity are still poorly understood. The fate and transport of contaminants in this highly human-altered system depend on a multitude of factors, but it is clear that contaminants arising from human activity well inland in the estuary system can have impacts very far afield, perhaps including coastal regions beyond the mouth of the estuary.

4

Governing Dynamics

4.1   Introduction

The currents in the urban ocean influence the physical, chemical, and biological processes that occur. They are the result of the fundamental governing equations describing urban ocean circulation. The governing equations originate with Newton's second law of motion, stating that the sum of the forces (F) on a parcel of water is equal to mass (m) of that parcel times its acceleration (a) and can be written as,

$$m \, \vec{a} = \sum \vec{F} \, . \tag{4.1}$$

Isaac Newton developed this law of motion (Equation 4.1) in 1687. We should be able to understand what he did, as the law has been with us for more than 330 years.

To discuss the forces and motions associated with the currents requires a coordinate system. Here we assume that the region of the Earth we are considering is small when compared to the size of the Earth itself. We can appeal to a rectilinear Cartesian coordinate system in which the Earth is assumed to be locally flat.

The convention we will use is: the x axis points east with u, the east–west velocity; the y axis points north with v, the north–south velocity; and the z axis points up with velocity w (0 at sea surface) as shown in Figure 4.1. Gravity points in the negative z direction. We need to consider all three directions because a parcel of water can move in the three directions under forces governed through Newton's second law of motion.

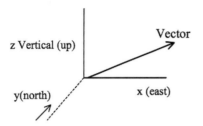

Figure 4.1 The Cartesian coordinate system used in this text.

4.2   Forces

The forces on a parcel of water are the pressure gradient, friction, Coriolis, and gravity. There are only these four forces, as written in Equation 4.2. Dividing Equation 4.1 by the mass and noting that the acceleration, a, is the time rate of change of velocity, Newton's second law for the ocean becomes,

$$m\frac{d\vec{v}}{dt} = \overrightarrow{coriolis} + \overrightarrow{press\ grad} + \overrightarrow{gravity} + \overrightarrow{friction}. \tag{4.2}$$

Equation 4.2 is called the equation of motion. It represents motion in three dimensions.

Let us consider the forces in Equation 4.2 one at a time. The easiest to understand is the gravitational force, where gravity, g, is 9.8 m/s². Fluid parcels are pulled vertically downwards in the z direction toward the Earth's center of mass by gravitational attraction. The magnitude of g changes slightly by about 5%; it is 9.78 m/s² at the equator and 9.83 m/s² at the poles. It varies slightly because the Earth is not perfectly round and not of uniform density.

Fluid parcels tend to flow from areas of high pressure to areas of low pressure with the velocity being proportional to the pressure gradient. The greater the pressure gradient, the faster the flow. The pressure gradient force is easy to visualize. It can be looked at in terms of the derivative $dPdx$ along the x axis. This is the pressure gradient. The derivative is simply stated along the axis as the pressure on the right minus the pressure on the left, $P_B - P_A$ divided by the distance between them, $X_B - X_A$. If P_B is a greater pressure than P_A, the pressure gradient force will be directed toward X_A. This pressure gradient is positive:

$$
\begin{array}{cccc}
x_A & \dfrac{dp}{dx} \simeq \dfrac{P_B - P_A}{x_B - x_A} & & x_B \\
P_A & & & P_B
\end{array}
$$

⬅━━━━━━━━━━━━━━━

leading to a velocity that is positive (pointing to the right, toward the increasing x axis). However, we know that if there's a pressure gradient pushing flow from high to low, that velocity should be in the opposite direction, i.e., negative. A minus sign is introduced to obtain the proper directionality. The pressure gradient force in three dimensions becomes

$$PGF_x = -\frac{1}{\rho}\frac{\partial P}{\partial x},$$ (4.3)

$$PGF_y = -\frac{1}{\rho}\frac{\partial P}{\partial y},$$ (4.4)

$$PGF_z = -\frac{1}{\rho}\frac{\partial P}{\partial z},$$ (4.5)

where ρ is the ocean water density.

The Coriolis force is arguably the most interesting force that governs how a fluid parcel would move. It incorporates the effect of the rotation of the Earth. It is very complicated, and it is possible that not many scientists understand its full complexity. Objects moving in air and in water on the Earth's surface are decoupled from the solid earth and move independently. Coriolis deflection is an apparent movement (to an observer viewing the Earth from space), due to the fact that the Earth's speed of rotation is slower at the poles than the equator.

For the purposes of this discussion, let us simplify the force as much as possible. If you take a ball and throw it up in the air, gravity brings it back to you. If you throw the ball up even higher it still returns to you. Now, if you throw it very, very high, what happens is the ball does not come back to you. Instead it lands somewhere near you. You quickly conclude that some force has pushed the ball off its intended path back to you. However, a person sitting on the moon watching you would come to a different conclusion. They would say you moved as the ball was in the air. You are on the Earth and the ball isn't. As the Earth turns, it takes you along with it, but the ball comes straight down.

Another way to look at the Coriolis force is to stand at the North Pole and launch a rocket toward the equator, as illustrated in Figure 4.2. As that rocket starts to move, the Earth underneath that rocket is turning. If you are sitting on that rocket, you assume that some force is pushing it since the trajectory you planned is going to miss the intended landing spot. In both examples, it was the Earth moving that caused the "missing force". The Coriolis force is an inertial force that acts on fluid parcels that are in motion relative to a rotating reference frame.

The Coriolis force causes fluid parcels to be pushed to the right in the Northern Hemisphere and to the left in the Southern Hemisphere. The force is proportional to the speed of the parcel and its latitude on Earth. The force vanishes if the parcel is at rest or located along the equator and is at a maximum at the poles.

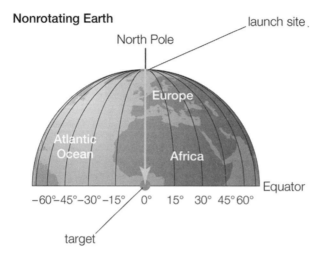

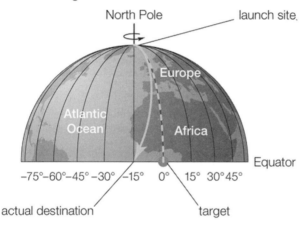

Figure 4.2 Path of a rocket launched toward the Equator from the North Pole showing that the projectile would land to the right of its true path (Reprinted with permission from the *Encyclopedia Britannica*, © 2008 by Encyclopedia Britannica, Inc.) (A black-and-white version of this figure appears in some formats. For the color version, please refer to the plate section.)

The most appropriate way to include the Coriolis force in the urban ocean is by adding an apparent force. In the coordinate system we are using, this force is written to a very useful degree of accuracy for the x component of flow as $F_c/m = 2\Omega v \sin(\varphi)$ and for the y component of flow as $F_c/m = -2\Omega u \sin(\varphi)$. Here $\Omega = 7.292 \times 10^{-5}$/s, the rotation rate of the Earth, and φ is the latitude. In the governing equations, the Coriolis force in the x and y directions can be written as

$$CF_x = fv, \tag{4.6}$$

$$CF_y = -fu. \tag{4.7}$$

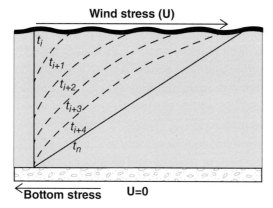

Figure 4.3 The effect of friction distributing the wind stress throughout the water column from times t_i to t_n.

where $f = 2\Omega\sin\varphi$ is the "Coriolis parameter" and ranges from -1.4×10^{-4}/s at the South Pole to $+1.4 \times 10^{-4}$/s at the North Pole. $\mathbf{f} \approx$ about 1×10^{-4}/s for mid-latitudes. There is no meaningful contribution to the motion of a fluid parcel by the vertical component of Coriolis force.

The Coriolis force is a large-scale process related to the rotation of the Earth. A measure of the importance of the effect of the Earth's rotation is the Rossby number (R_o), i.e., the ratio of the field acceleration to the Coriolis force:

$$R_o = \frac{U^2/L}{f_o U} = \frac{U}{(f_o L)}. \tag{4.8}$$

The smaller the value of R_o, the more important the effects of the Earth's rotation on the currents. In the urban ocean $f_o \sim 10^{-4}$ s, $L \sim 50 \times 10^3$ m, and $U \sim 1$ m/s, suggesting that $R_o \sim 0.2$. While not as small as in the atmosphere, the Coriolis force still plays a role, even in a small bay. In the Hudson River, which is fairly narrow (5 km wide), the saltiest water is found on the right side looking from the ocean end inwards.

The role of friction, the last of the forces we seek to understand, is to transfer the forces on the edges of a fluid into the interior of the fluid itself. Consider a wind blowing on the surface of a body of water, as shown in Figure 4.3.

The bottom is fixed, and the surface is accelerated by applying a wind stress that acts from left to right. The surface will be accelerated to some steady velocity, and the fluid between the surface and bottom will be set in motion. A steady velocity is achieved when the applied wind stress is balanced by a resisting stress (shown as an equal but opposite stress applied by the stationary bottom). As the surface begins to accelerate, the velocity of the fluid parcels in contact with the surface is equal to the velocity at the surface (a no slip condition exists between the surface wind and the fluid). Fluid parcels in contact with those at the surface will be accelerated due to frictional attraction between them and so on through the column of fluid. The mixing by the eddy viscosity of the fluid, K_M, results in layers of fluid that are increasingly further from the surface being set in motion. The bottom and fluid parcels near to it are stationary (zero velocity at the bottom) so that eventually the velocity will vary from zero at the bottom to U_{STEADY} at the surface, which is equal to the steady velocity of the surface. The velocity gradient (the rate of change in velocity between surface and bottom; du/dz) will be constant, and the velocity will increase linearly with a constant K_M from zero at the bottom to U_{steady} at the surface. Steady velocity is achieved when the resisting stress (the bottom stress shown applied by the bottom) is equal but opposite to the wind stress applied to the surface.

The stress transfers momentum (mass times velocity) through the fluid to maintain the linear velocity profile. The relationship between the frictional force acting on the U velocity, the fluid eddy viscosity, and the velocity gradient is given by

$$\frac{\partial}{\partial z}\left(\frac{K_M \partial u}{\partial z}\right). \tag{4.9}$$

Extending this idea to three dimensions we can write the frictional force contribution to Equation 1 as

$$x: \frac{\partial}{\partial x}\left(\frac{A_M \partial u}{\partial x}\right) + \frac{\partial}{\partial y}\left(\frac{A_M \partial u}{\partial y}\right) + \frac{\partial}{\partial z}\left(\frac{K_M \partial u}{\partial z}\right), \tag{4.10}$$

$$y: \frac{\partial}{\partial x}\left(\frac{A_M \partial v}{\partial x}\right) + \frac{\partial}{\partial y}\left(\frac{A_M \partial v}{\partial y}\right) + \frac{\partial}{\partial z}\left(\frac{K_M \partial v}{\partial z}\right), \tag{4.11}$$

$$z: \frac{\partial}{\partial x}\left(\frac{A_M \partial w}{\partial x}\right) + \frac{\partial}{\partial y}\left(\frac{A_M \partial w}{\partial y}\right) + \frac{\partial}{\partial z}\left(\frac{K_M \partial w}{\partial z}\right). \tag{4.12}$$

Here A_M is the horizontal eddy viscosity or horizontal mixing coefficient, and K_M, the vertical eddy viscosity or vertical mixing coefficient. Values for these coefficients will be discussed in the next chapter.

The presence of a turbulent flow is measured by a non-dimensional number, the Reynolds number R_e,

$$R_e = \frac{uL}{v}, \tag{4.13}$$

where U is a typical velocity of the flow, L is a typical length describing the flow, and v the molecular viscosity of water. The Reynolds number is named after Osborne

Reynolds (1842–1912) who conducted experiments in the late nineteenth century to understand turbulence. He found that when the speed of water flow in a tube was low, the flow was smooth without any turbulence. This is called laminar flow. At higher speeds, the flow became irregular and turbulent. The transition occurred at $Re = UL/v \approx 2{,}000$. As the Reynolds number increases above a critical value of 4000, the flow becomes more and more turbulent. With a molecular viscosity $(v) = 1$ $E{-}6$ m^2/s at 20°C and with typical values in the urban ocean of $U = 1$ m/s and $L = 10$ km we compute that $Re = 10^9$. We see that the urban ocean is always turbulent.

Turbulence in the urban ocean is associated with velocity gradients which leads to, for example, wind-induced mixing in the surface layers as discussed above. Tidal motions create velocity gradients near the bottom, particularly in the shallower waters of the urban ocean. Enhanced turbulence and mixing is created by vertical gradients of density where the surface density is higher than that below it. Because the ocean mostly has stable stratification, vertical mixing must work against this stratification. Vertical mixing requires more energy than horizontal mixing. As a result, horizontal mixing along surfaces of constant density is much greater than vertical mixing across surfaces of constant density. The latter, however, is very important because it changes the vertical structure of the water column, and it controls to a large extent the rate at which deeper water eventually reaches the surface.

4.3   Lagrangian and Eulerian Perspectives

Newton's second law of motion describes the motion of a parcel of water. It is called the Lagrangian description of fluid flow as it tracks the position and velocity of individual fluid parcels as they move about and determines how the fluid properties of these particles change as a function of time. This description is difficult to use for practical flow analysis because fluids are composed of billions of such parcels, that is, a fluid is a continuum. In the Eulerian perspective, the fluid motion is given by completely describing the necessary properties as a function of space and time. We obtain information about the flow by noting what happens at fixed points. If we have enough information, we can obtain Eulerian from Lagrangian or vice versa. The difference between the two descriptions is made clearer by imagining a person standing beside a river, measuring its properties. In the Lagrangian approach, the person throws probes into the water that move downstream with the water. In the Eulerian approach, the person anchors probes at a fixed location in the water. Figure 4.4 compares the Lagrangian to the Eulerian approach in terms of how water flows around a pier.

In oceanography, analysis is easier in terms of field (Eulerian) properties, $P(x,y,z,t)$. It is well suited for the formulation of initial boundary value problems

(a) Lagrangian

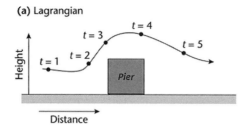

(b) Eulerian

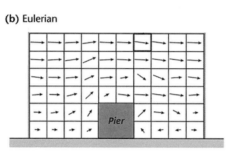

$$t = \{1, 2, 3, \ldots\}$$

Figure 4.4 Conceptual diagram illustrating the (a) Lagrangian and (b) Eulerian approaches to understanding flow patterns in the vicinity of a pier. The position of the probe at various times is denoted by $t = 1$, $t = 2$, etc (adapted from Oke et al, 2017).

that are most common in the urban ocean. The relationship between the two descriptions is

$$\vec{a} = \frac{D\vec{u}}{Dt} = \frac{\partial \vec{u}}{\partial t} + \left(\vec{u} \cdot \vec{\nabla}\right) \vec{u} \tag{4.14}$$

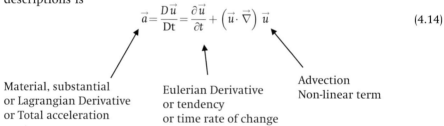

Material, substantial or Lagrangian Derivative or Total acceleration

Eulerian Derivative or tendency or time rate of change

Advection Non-linear term

The Eulerian derivative is also called the local acceleration or tendency. In component form the acceleration becomes

$$a_x = u\frac{\partial u}{\partial x} + v\frac{\partial u}{\partial y} + w\frac{\partial u}{\partial z} + \frac{\partial u}{\partial t}, \tag{4.15}$$

$$a_y = u\frac{\partial v}{\partial x} + v\frac{\partial v}{\partial y} + w\frac{\partial v}{\partial z} + \frac{\partial v}{\partial t}, \tag{4.16}$$

$$a_z = u\frac{\partial w}{\partial x} + v\frac{\partial w}{\partial y} + w\frac{\partial w}{\partial z} + \frac{\partial w}{\partial t}. \tag{4.17}$$

The adjectival term is "nonlinear" because the velocities occur as products of the velocity components and their various spatial derivatives. Because of these non-linear terms, a small perturbation may grow into a large fluctuation leading to turbulence that occurs whenever the fluctuations are sufficiently large compared to the frictional terms, which tend to remove or smooth out velocity differences. This measurement of "large enough" is given by the Reynolds number in Equation 4.13.

4.4   Momentum Equations

Putting all the forces together we get our first equation set:

x-momentum

$$\frac{\partial u}{\partial t} + u\frac{\partial u}{\partial x} + v\frac{\partial u}{\partial y} + w\frac{\partial u}{\partial z} = -\frac{1}{\rho_0}\frac{\partial p}{\partial x} + fv + A_H\nabla_H^2 u + \frac{\partial}{\partial z}K_M\frac{\partial u}{\partial z}, \tag{4.18}$$

y-momentum

$$\frac{\partial v}{\partial t} + u\frac{\partial v}{\partial x} + v\frac{\partial v}{\partial y} + w\frac{\partial v}{\partial z} = -\frac{1}{\rho_0}\frac{\partial p}{\partial y} - fu + A_H\nabla_H^2 v + \frac{\partial}{\partial z}K_M\frac{\partial v}{\partial z}, \tag{4.19}$$

z-momentum

$$\rho g = -\frac{\partial P}{\partial z}. \tag{4.20}$$

The z-momentum equation reduces to become the hydrostatic equation because the vertical acceleration terms and the frictional terms are unimportant in the ocean. Equation 4.20 says that the pressure at any depth in the water column is due to the weight of all the water above that depth.

Vertically integrating the hydrostatic equation from a depth z to the free surface yields

$$P(x,y,z,t) = P_{\text{atm}} + g\rho_0\eta + g\int_z^0 \rho\left(x,y,z',t\right)dz', \tag{4.21}$$

where η is the distance from the datum $Z = 0$ to the surface. Inserting the pressure into the pressure gradient term in Equation 4.20 three contributions to the pressure gradient emerge. They are the "barometric", the "barotropic", and the "baroclinic" contributions. The contributions are written for the x component as

$$-\frac{1}{\rho}\frac{\partial P}{\partial x} = \frac{1}{\rho}\left[-\frac{\partial P_a}{\partial x} - \rho g\frac{\partial \eta}{\partial x} - g\int_z^0 \frac{\partial \rho}{\partial x}dz\right], \tag{4.22}$$

barometric   barotropic   baroclinic

where P_a is the atmospheric pressure and η is the free surface height. Only the baroclinic contribution is a function of depth, z. The first term on the right-hand side is the atmospheric component. If a high-pressure system is coming, you will experience a pressure gradient from the atmosphere pushing or pulling on top of the ocean. The barotropic component is about the water level changing horizontally. If the ocean is high on one side, low on the other, perhaps a tsunami is coming or a storm surge, as that change in water level from place to place generates a pressure gradient. The baroclinic component is related to how density changes. The gradient of the density is integrated from a depth z to the surface. It has depth dependence in it. This term is what drives the estuarine circulation described in the previous chapter.

Each term has a name and it is important to know them for the analysis to follow in subsequent chapters. The names are:

Total   Pressure   Coriolis   Horizontal
Acceleration   Gradient   Force   Friction
                                                                 Vertical
                                                                 Friction

$\diagdown$ $\diagdown$ $\diagup$ $\diagup$ $\diagup$

$$\frac{Du}{Dt} = -\frac{1}{\rho_0}\frac{\partial p}{\partial x} + fv + A_H \nabla_H^2 u + \frac{\partial}{\partial z} K_M \frac{\partial u}{\partial z} \tag{4.23}$$

$$\frac{Dv}{Dt} = -\frac{1}{\rho_0}\frac{\partial p}{\partial y} - fu + A_H \nabla_H^2 v + \frac{\partial}{\partial z} K_M \frac{\partial v}{\partial z} \tag{4.24}$$

$$0 = -\frac{1}{\rho}\frac{\partial P}{\partial z} - g \qquad\text{——— Gravity} \tag{4.25}$$

where, as we have seen, the total acceleration is composed of two terms, the local acceleration or tendency and the field acceleration.

5

Mass, Salt, and Temperature "Conservation"

5.1 Introduction

The conservation of volume, salt and freshwater, and heat energy (temperature) is the foundation for understanding how the urban ocean works. The conservation of volume principle (often called the equation of continuity) is a result of the compressibility of water being so small. If water is flowing into a closed, control volume at a certain rate, it must be flowing out somewhere else at the same rate or the level in the control volume must increase. The transport of salt and heat, or any scalar for that matter, can also occur via turbulence mixing. This leads naturally to the concepts of horizontal and vertical eddy diffusivity and builds on the notion of horizontal and vertical eddy viscosity discussed in the previous chapter.

5.2 Conservation of Mass

In the ocean, it is possible to assume that the water is incompressible, that is, its volume cannot change no matter how much it is squeezed. Consider the control volume of water in Figure 5.1.

In the x direction, the net volume of flow in is

$$u\delta y\delta z \tag{5.1}$$

and the volume of flow out is

$$\left[u + \frac{\partial u}{\partial x}\delta x\right]\delta y\delta z. \tag{5.2}$$

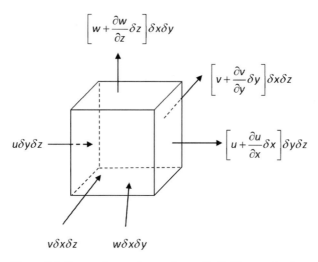

Figure 5.1 Fluxes of water into and out of a fluid control volume.

The complete set of volume flows as sketched in Figure 5.1 is

$$u\delta y\delta z - \left[u + \frac{\partial u}{\partial x}\delta x\right]\delta y\delta z + v\delta x\delta z - \left[v + \frac{\partial v}{\partial y}\delta y\right]\delta x\delta z + w\delta x\delta y - \left[w + \frac{\partial w}{\partial z}\delta z\right]\delta x\delta y$$

$$= -\frac{\partial u}{\partial x}\delta x\delta y\delta z - \frac{\partial v}{\partial y}\delta x\delta y\delta z - \frac{\partial w}{\partial z}\delta x\delta y\delta z = 0. \tag{5.3}$$

This volume summation says that whatever flows go into the control volume must come out somewhere. Volume cannot be "packed" in or "thinned out". What we see in an urban ocean control volume is that if more (or less) water goes into the volume than comes out, the free surface will rise (drop). Now dividing through by the volume, *dxdydz*, we obtain the conservation of mass or continuity equation,

$$\frac{\partial u}{\partial x} + \frac{\partial v}{\partial y} + \frac{\partial w}{\partial z} = 0. \tag{5.4}$$

Suppose an incompressible flow with velocity v_{in} = 0.10 m/s from the continental shelf is flowing into a river of depth H_{in} = 20 m and width L = 1 km, as in Figure 5.2. Upstream there is a section of depth H_{out} = 5 m. What is the upstream velocity v_{out} of this flow?

In general, the continuity equation cannot be used by itself to solve for flow field; however, it can be used to determine if the velocity field is incompressible or to find the missing velocity component if you have two of the components.

For example, given an incompressible velocity field with components $u = ax^2 - ay^2$ and $w = b$ where a and b are constants, determine the velocity component v. Taking the derivatives

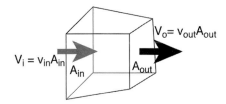

Figure 5.2 Flow balance at the junction between the continental shelf and a river.

V_i = input volume flow rate = $v_{in} A_{in}$ = 0.10 m/s × 10^3 m × 20 m = 2,000 m³/s

V_o = output volume flow rate = $v_{out} A_{out}$ = v_{out} × 10^3 m × 5 m = v_{out} × 5,000 m²

Given conservation of mass, $V_i = V_o \geq v_{out} = V_i/A_{out}$ = 0.40 m/s

$$\frac{\partial u}{\partial x} = 2ax, \tag{5.5}$$

$$\frac{\partial w}{\partial z} = 0, \tag{5.6}$$

$$\frac{\partial v}{\partial y} = ? \tag{5.7}$$

$$2ax + \frac{\partial v}{\partial y} + 0 = 0, \tag{5.8}$$

and then integrating with respect to y, the answer is

$$v = -2axy + c (c \text{ is a constant}). \tag{5.9}$$

5.3   Conservation of Heat and Salt

The ideas applied above to the mass balance of a fixed volume can also be applied to other scalar quantities. Here the temperature or salinity inside the control volume can change depending on the rate at which the scalar quantity is transported across the six sides of the control volume. The temperature can be changed by current advection, surface heating and cooling, and by turbulent mixing. Salinity can be changed by current advection, evaporation, precipitation/runoff, brine rejection during ice formation, and turbulent mixing. Density as we have learned in Chapter 2 is related to temperature and salinity through the equation of state.

The conservation of heat/temperature is written as

$$\frac{DT}{Dt} = \frac{\partial}{\partial x}\left[A_H \frac{\partial T}{\partial x}\right] + \frac{\partial}{\partial y}\left[A_H \frac{\partial T}{\partial y}\right] + \frac{\partial}{\partial z}\left[K_z \frac{\partial T}{\partial z}\right] \tag{5.10}$$

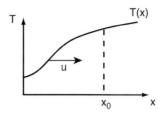

Figure 5.3 Schematic showing how the temperature at a point X_0 can change from two different mechanisms.

and the conservation of salinity is written as

$$\frac{DS}{Dt} = \frac{\partial}{\partial x}\left[A_H \frac{\partial S}{\partial x}\right] + \frac{\partial}{\partial y}\left[A_H \frac{\partial S}{\partial y}\right] + \frac{\partial}{\partial z}\left[K_z \frac{\partial S}{\partial z}\right], \tag{5.11}$$

where the material derivative as before is

$$\frac{D}{Dt} = u\frac{\partial}{\partial x} + v\frac{\partial}{\partial y} + w\frac{\partial}{\partial z} + \frac{\partial}{\partial t}. \tag{5.12}$$

The material derivative (see Chapter 4) can be visuaized by considering the temperature equation in one dimension, $T(x,t)$,

$$\frac{DT}{Dt} = \frac{\partial T}{\partial t} + u\frac{\partial T}{\partial x}. \tag{5.13}$$

Using the setting shown in Figure 5.3, at given point x_0, a change in temperature can be caused by two different mechanisms:

1. The temperature of the local fluid particle changes in time (due to internal sources, for example, heat conduction, radioactive heating, etc.). In Equation (5.13),

$$\frac{\partial T}{\partial t} \neq 0. \tag{5.14}$$

2. All fluid particles keep their temperature, but the velocity u brings a new particle to x_0, which has a different temperature. In Equation (5.13),

$$u\frac{\partial T}{\partial x} \neq 0. \tag{5.15}$$

Consider a steady flow of water flowing downstream in an estuary past an area where a power plant is introducing heated water to the estuary as shown in Figure 5.4.

The temperature in the estuary is changing as $T = 20°C + (1°C/km)\,X$, where X is the distance along the estuary. The velocity in the estuary is increasing as $u = 1$ m/s $+ (5 \times 10^{-2}$m/s per km$)\,X$. Now we can determine the rate of change

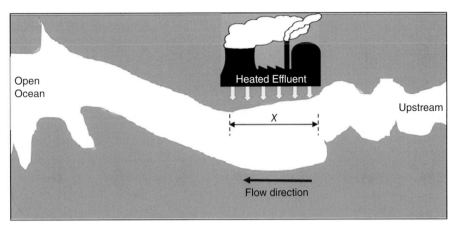

Figure 5.4 Section of an estuary with an adjoining power plant spanning a distance X.

of the temperature as the flow passes a point 10 km from where the heat first enters.

From Equation (5.13), we have,

$$\frac{DT}{Dt} = 0 + \left[1 \text{ m/s} + \left(5 \times 10^{-2} \text{m/s/km}\right)(10 \text{ km})\right](1 \text{ C/km}) = 1.5 \times 10^{-3} \text{C/s}.$$

Expanding the material derivative given by Equation 5.12, the conservation equations for temperature (Equation 5.10) and salinity (Equation 5.11) can be written as

$$\frac{\partial T}{\partial t} + u\frac{\partial T}{\partial x} + v\frac{\partial T}{\partial y} + w\frac{\partial T}{\partial z} = \frac{\partial}{\partial x}\left[A_H\frac{\partial T}{\partial x}\right] + \frac{\partial}{\partial y}\left[A_H\frac{\partial T}{\partial y}\right] + \frac{\partial}{\partial z}\left[K_z\frac{\partial T}{\partial z}\right], \tag{5.16}$$

$$\frac{\partial S}{\partial t} + u\frac{\partial S}{\partial x} + v\frac{\partial S}{\partial y} + w\frac{\partial S}{\partial z} = \frac{\partial}{\partial x}\left[A_H\frac{\partial S}{\partial x}\right] + \frac{\partial}{\partial y}\left[A_H\frac{\partial S}{\partial y}\right] + \frac{\partial}{\partial z}\left[K_z\frac{\partial S}{\partial z}\right]. \tag{5.17}$$

The first terms on the left-hand side of Equations 5.16 and 5.17 are the tendency terms. They express how heat (temperature) and salinity change in time. The remaining terms on the left side are called the advection terms. They express carriage of heat (temperature) and salinity from place to place. Horizontal and vertical friction are captured by the terms on the right-hand side of the two equations.

The turbulence mixing in Equations 5.16 and 5.17 is represented by the horizontal A_H and A_H and vertical K_z eddy diffusivities, also known as the mixing coefficients. The problem is, the determination of the appropriate turbulence mixing coefficients" A_M, and K_M for momentum and A_H and A_H and K_z for temperature and salinity or any other scalar is a rather uncertain art. Estimates for these are provided in Figure 5.5. Unlike molecular diffusivities, K_M and K_z are approximately equal. Semi-empirical, turbulence closure techniques have had

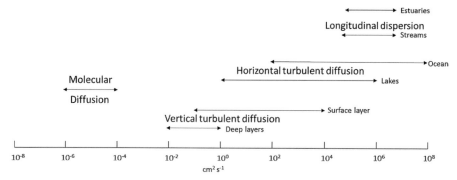

Figure 5.5 Estimates for the horizontal and vertical mixing coefficients in the environment.

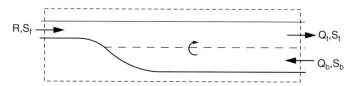

Figure 5.6 Flow balance in a highly stratified two-layered estuary.

some success in alleviating the problem (Mellor and Yamada, 1982). In this book, we will not delve into these techniques.

5.4 Knudsen Hydrographical "Theorem"

In stratified estuaries, salt transport is a balance between the horizontal and vertical current advection of salt. Knudsen used that to develop a hydrographical theorem that provides an easy way to determine the mean flow of water, Q, in stratified estuaries. Consider the situation shown in Figure 5.6, where S_t is the mean salinity of the top layer, S_b the mean salinity of the bottom layer, and Q_t is the flow (actually volume flow rate in m^3/s) of the top layer exiting the estuary and Q_b is the flow of the bottom layer entering the estuary. The river flow that is typically known entering this estuary is R and its salinity is S_f mostly 0, but not always. The mean in this situation is an average over the period in which the river flow is known.

Conservation of volume states that $Q_t + R = Q_b$, that is, the volume of water that enters the estuary equals the volume of water that leaves it. The water entering the bottom layer will eventually work its way into the top layer for its exit to the ocean. Similarly, conservation of salt requires that the amount of salt entering the channel has to be balanced by the amount of salt leaving it: $Q_t\, S_t = Q_b\, S_b$.

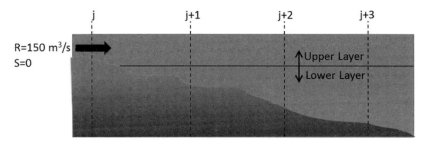

Figure 5.7 Schematic of a highly stratified estuary where salinity observations were taken at four locations, j to $j + 3$.

The volume balance and the salt balance provide two equations for the two unknowns Q_t and Q_b. Solving for them yields

$$Q_t = \frac{R}{1 - S_t/S_b}, \tag{5.18}$$

$$Q_b = \frac{R}{S_b/S_t - 1}. \tag{5.19}$$

Examining these equations suggests that the mean flows in a highly stratified estuary will be much less than those of a partially mixed estuary as the ratio of S_t/S_b is greater in the highly stratified case. And for a vertically mixed estaury the equations break down because the demoninator in Equations 5.18 and 5.19 approach 0, suggesting an infinite set of flows.

We can generalize this result by looking at the volume and salt balance for individual sections of the estuary. Assume that observations of salinity were taken along an estuary at a number of stations and consider the situation between two neighboring stations j and $j - 1$. Following our conservation of volume and salt principles of "what goes in, has to go out", we find for the volume fluxes

$$Q_i^{upper} - Q_i^{lower} = Q_{i-1}^{upper} - Q_{i-1}^{lower} \tag{5.20}$$

and for the salt fluxes

$$\left(Q_i^{upper} S_i^{upper}\right) - \left(Q_i^{lower} S_i^{lower}\right) = \left(Q_{i-1}^{upper} S_{i-1}^{upper}\right) - \left(Q_{i-1}^{lower} S_{i-1}^{lower}\right). \tag{5.21}$$

Once you know the solution (volume transports) at one location $j-1$ you can obtain the solution at another location j.

Consider the highly stratified estuary shown in Figure 5.7. Assume that observations of salinity, S_t and S_b, were taken along the estuary at a number of stations (see Table 5.1). Find Q_t and Q_b at each station using Knudsen's

Table 5.1 *Salinity along the axis of an estuary along with the flows that result*

	j	$j + 1$	$j + 2$	$j + 3$
S_t (ppt)	0	10	20	30
S_b (ppt)	–	35	35	35
Q_t (m^3/s)	150	210	350	1050
Q_b (m^3/s)	–	60	200	900

hydrographical theorem. Assume there is an inflowing river with a freshwater discharge of 150 m^3/s.

It is important to note how continuity works here. The Q_t at $j + 1$ is 210 m^3/s and the flow at j is 150 m^3/s. The reason this occurred was because the transport in the bottom layer at $J + 1$ flowed vertically upwards to join the flow at j and all flowed outwards in the top layer.

The core assumption of the Knudsen formula is that all sea water that enters the estuary leaves it in the upper layer after complete mixing with fresh water. In reality, the Knudsen theorem overestimates the volume transported in partially stratified estuaries. In partially stratified estuaries salt is transported both upward and downward by turbulent tidal mixing. While there is much exchange between the two layers, the net upward transport of water is much less than in a highly stratified estuary. It follows that Knudsen's hydrographical theorem can be used to get an upper estimate of upper and lower transport in estuaries but that the result must be interpreted with caution if the estuary is not of the highly stratified type.

5.5   Residence Time

And finally, we are going to develop a framework to quantify the residence time of an estuary whose flows can be determined by Knudsen's theorem. This is especially important in those estuaries where it has been a convenient place to dispose of unwanted materials and pollutants. The residence time is most useful in determining levels of contamination and concentrations of nutrients in a water body and their potential effects on estuarine organisms. It is commonly used in producing comparative assessments of how estuaries respond to human use. The estuarine residence time is the time it takes to completely exchange its water. A short residence time suggests a rapid removal of pollutants and debris. The importance of residence time for nontoxic substances is that shorter residence times reduce the risk of primary or secondary effects occurring. For example, the risk of deoxygenation of the water column is reduced if the oxygen-depleting substances are better mixed in the water column and are washed out of an estuary quickly. Similarly, the risk of the secondary effects of eutrophication are reduced if the phytoplankton are rapidly removed from an estuary and do not die off causing oxygen depletion. For toxic substances, shorter residence times result in reduced contamination of the water column and sediments and reduced exposure periods for organisms in the water body. The definition of residence time is

$$t_F = \frac{Volume}{Q_t} \tag{5.22}$$

or

$$t_F = \frac{V}{Q_{top}} = \frac{V}{\left(\dfrac{R}{\left(1 - \dfrac{S_{top}}{S_{bottom}}\right)}\right)}. \tag{5.23}$$

The residence time is the volume of the estuary upstream of where the salinities are measured divided by discharge of the top layer (outflowing layer). Estuaries with long residence times flush slowly. They can be expected to have different characteristics than those that flush quickly, especially with regard to the rate of introduction of water and associated materials, such as nutrients and sediments, and the extent of within-estuary processing of those materials. For example, the extent of nutrient retention within estuaries versus transport through them has been related to various measures of mixing time in North Atlantic estuaries (Nixon et al., 1996), and the growth rate of phytoplankton relative to the rate of physical flushing of water can determine if blooms are likely to occur.

Consider Barnegat Bay–Little Egg Harbor, New Jersey, a back-barrier estuary that is subject to anthropogenic pressures including nutrient loading, eutrophication, and subsequent declines in water quality. Mean residence time is 13 days. The Mersey Estuary is one of the largest and most heavily polluted estuaries in the UK, with a catchment area of about 5000 km^2 over the northwestern part of England that includes the major contributions of the cities of Liverpool and Manchester. It has a relatively short average residence time for a large estuary, ranging from 17 to 104 hours (i.e., from less than 1 day to about 4 days). Concerns have grown over the increase of nutrients and pollutants discharged into the Chesapeake Bay in recent years. The mean retention time of the entire Chesapeake Bay system ranges from 110 to 264 days, with an average value of 180 days. The retention time was greater in the bottom layers than in the surface layers due to the persistent stratification and estuarine circulation.

6

Water Level Changes

6.1   Introduction

The rise and fall of the ocean water surface along the world's coastlines has a profound effect on the physical, chemical, and biological systems and their interactions with the land and with human populations. The tide, a seemingly simple phenomenon driven by the gravitational attraction of the sun, the moon, and the other astronomical bodies, is actually among the most complex influences on the coastal environment. Its behavior is governed by factors that include coastline geometry, bottom topography, and interactions with coastal rivers and bays. The tide drives the fluxes of salinity, temperature, and nutrients that sustain life along the coast and often hundreds of miles inland. It produces wetting and drying cycles that define unique ecosystems such as wetlands and have for millennia helped define the location of human habitats.

Although a dominant and very predictable contribution to water level changes in coastal regions, the tide is by no means the only contributor. In the context of urban coastal regions, several other, less persistent, less predictable, but potentially far more impactful water level phenomena must be considered. These include exceedingly rare but high-consequence events such as tsunamis and more frequent, potentially very impactful events such as storm surges caused by tropical and extratropical cyclones. Large-scale currents such as the Gulf Stream in the North Atlantic and the Kuriosho Current in the North Pacific, as well as powerful but more localized coastal flows such as upwelling and downwelling, can also produce significant changes in the position of the sea surface. Superimposed on all of these water level changes is the steady but spatially highly variable phenomenon of sea level rise.

This chapter will provide a thorough discussion of the physics of water level changes and the influence of these changes on landforms and ecosystems, and, in turn, on human populations. Although much of this discussion will necessarily address the coastal hazards associated with extreme water levels, we will also consider the emerging opportunities associated with this dominant feature of the coastal environment, including renewable energy extraction.

6.2   The Tide

The tide is a wave (yes, the true "tidal wave") that propagates along the world's coastlines with highly predictable height and frequency. The tide is caused primarily by the gravitational force of attraction of the moon and the sun. The gravitational force of attraction of the other astronomical bodies also contribute, but to a much lesser extent.

Examining first the effect of the moon, as shown in Figure 6.1, the strength of the force due to gravity is highest on the side of the Earth closest to the moon, resulting in a tidal "bulge" at the location in the ocean nearest to the moon, and at the location on the opposite side of the Earth. But why?

We know that the force due to gravity is proportional to the mass of the two objects being considered, and inversely proportional to the distance between the two objects *squared*. The sun is much larger than the moon, with a mass 27 million times greater, and it is located 390 times further away from the Earth. But it turns out that the tide-generating effect of the sun is only approximately 46% of that due to the moon. Also, if we did the calculations, we would find that the attractive force due to gravity of either the sun or the moon is not sufficient to produce the tidal bulge observed in the ocean. The reason is that the

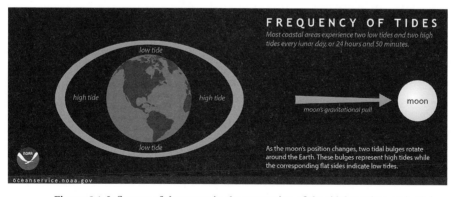

Figure 6.1 Influence of the moon in the generation of the tidal maxima and minima (NOAA, 2017a).

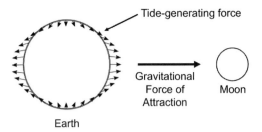

Figure 6.2 Pattern of the tide-generating force associated with the influence of the moon.

tide-generating force due to the gravitational force of attraction is actually associated with the *gradient* in the gravitational force along the surface of the ocean. As already stated, the force due to gravity is greatest at the point nearest the moon (or sun), and lowest at the points 90 degrees from this location. This gradient in the gravitational force of attraction on all sides of the ocean surface results in a force – the tide-generating force – that is directed toward the moon (sun) at the location nearest the moon (sun) and directed inward at points located 90 degrees from this location, with vectors as depicted in Figure 6.2. The tide-generating force is directed in the opposite direction on the other side of the earth, as shown in Figure 6.2.

The tidal bulges appear at the location in the ocean nearest the moon and the point on the opposite side of the Earth because the water is displaced toward these locations, much as water flows downhill, from high pressure to low pressure. The fact that the tide-generating force is associated with the gradient in the force due to gravity means that it is in fact proportional to the distance between the objects *cubed*. This gives rise to the pattern exhibited in Figure 6.2. Importantly, in answer to our earlier question, this is the reason why the sun exerts a smaller influence on the tide than the moon despite its much larger mass. We will mention here that the pattern shown in Figure 6.2 is the same if we chose to show the vector sum of the gravitational forces of attraction due to the Earth and moon, and the centrifugal force, when the analysis is done in a rotating coordinate system.

As one would expect, the position of the bulge, or tidal wave crest, moves as the Earth revolves on its axis. Since the moon is also revolving around the Earth, the combined motion results in a "lunar day" of roughly 25 hours (actually 24 hours, 48.8 minutes), meaning the crest (highest tide) occurs approximately 1 hour later each day. Since the moon revolves around the Earth with a roughly 1-month cycle, its position relative to a point in space repeats approximately once per month. Approximately twice per month at any one point on Earth and on the point on the opposite side of the Earth, the additive effects of the moon and the

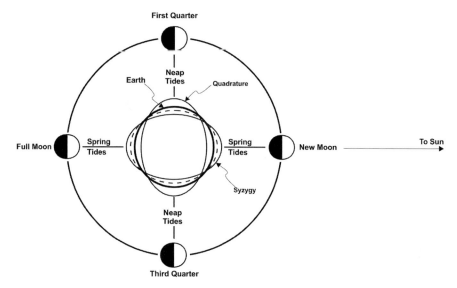

Figure 6.3 Looking down on the North Pole, we see the tide-generating force envelopes produced by the moon (solid ellipses) and the sun (dashed ellipses) at the positions of new and full moon (syzygy), and at the positions of first and third quarter (quadrature) (Gill and Schultz, 2001).

sun produce the maximum height of the tidal wave crest, often referred to as the "Spring Tide". When the moon and the sun are exactly out of alignment in their position relative to a location on Earth, the height of the tidal wave crest is reduced to its minimum, also referred to as the "Neap Tide". This is illustrated in Figure 6.3.

To complicate matters a bit more, we must remember that the Earth–moon system also revolves around the sun, with a 1-year cycle. Since the distance between the sun and any point on Earth changes as this revolution occurs, we experience cold and warm seasons. We also experience additional variability in the tidal wave crest height across the months. Over longer periods of time, the additional contributions to the tidal wave crest height due to the gravitational attraction of the planets result in additional small variations in the tide elevations. We can describe the various contributions to the tide in simple terms as a combination of individual waves, each associated with the object (moon, sun, etc.) that is responsible for that particular component. We call these "harmonic constituents". As we have already said, the "lunar day" is 24 hours, 48.8 minutes long, meaning that we experience two maximum lunar tides and two minimum lunar tides approximately every 24 hours, 50 minutes (recall the two bulges of water at any given moment in time). Therefore, we can express the lunar tide as a wave with a period (time between successive maxima or minima)

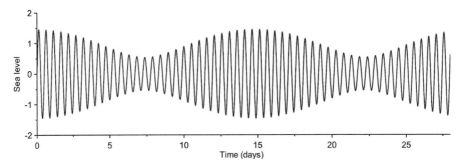

Figure 6.4 The waveform that results from the combination of a wave with 12-hour period (the S$_2$ constituent) and a wave with 12.42-hour period (the M$_2$ constituent).

of 12 hours, 25 minutes, or 12.42 hours. Of course, the solar tide has a period of 12 hours. The lunar component (referred to as the lunar "constituent") of the tide is called the "principal lunar" constituent and is referred to as the "M$_2$" tide. The solar constituent of the tide is called the "principal solar" constituent and is referred to as the "S$_2$" tide. They are both referred to as "semidiurnal" tidal constituents, meaning that each contributes to two high tides and two low tides per day.

What we have described to this point is a highly simplified description of the water level changes associated with the tide. It is easy to see that the reality is quite complex. We have only to examine Figure 6.4 to see that the combined influence of the lunar and solar tides is a complex combination of two different waves, each associated with a different orbit. The summation of these waves produces a complex but repeating pattern with a period equal to 14.8 days. We already mentioned this earlier in the context of the Spring and Neap tides, which occur twice per lunar cycle (approximately 1 month).

Figure 6.5 illustrates the water level record at the Atlantic Ocean coastline of Duck, North Carolina. Note that the pattern of the tide over this 1-month record looks nothing like the pattern illustrated in Figure 6.4! What is causing this?

For one part of the explanation, consider that the moon's orbit around the Earth is not a cirlce; it is an ellipse. This means that the elevation of successive Spring and Neap tides are not going to be the same. Mathematically, this can be included in tidal predictions as another "semidiurnal" consituent, the N$_2$ constituent. Now consider the fact that the Earth's rotational axis is tilted. This causes the tidal bulges on each side of the Earth to also be "tilted" relative to the equator. As a result, the two high tides and two low tides per day are not equal. This effect is obviously seen in both the lunar and solar constituents of the tide. Since this effect results in a "highest high" tide and a "lowest low" tide each day, the resulting constituents are "diurnal" constituents, with periods of

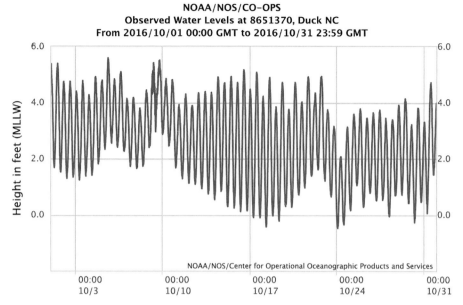

Figure 6.5 Water level record at Duck, North Carolina (NOAA, 2017b).

approximately 24 hours, and are labeled the O_1 (lunar diurnal), P_1 (solar diurnal), and K_1 (luni-solar diurnal) constituents.

Figure 6.6 illustrates the sea level pattern (relative height and length) for each of the six tidal constituents mentioned here and the resulting combined pattern when these constituents are summed together to form one wave.

Depending on the relative height of each of these six major constituents of the tide, the overall tidal sea level pattern can vary significantly from one location to another. Some may be dominated by the diurnal constituents and therefore have a single high tide and a single low tide each day; others may be dominated by the semidiurnal constituents and therefore have two prominent high tides and two prominent low tides each day; and still others may have a mix of influences and so have two high tides and two low tides per day, but with a very pronounced high tide and a very pronounced low tide each day. These three tide types are illustrated in Figure 6.7.

What determines which tide type will exist at a particular location? Further, what is responsible for the very large variations in the tidal "range" (the difference in water level between low tide and high tide) at locations around the Earth? For example, we list in Table 6.1 a few of the locations where the mean tidal range is among the highest in the world. What can cause such ranges when, for example, the mean tide range at Duck, North Carolina is 3.22 feet (NOAA, 2018b)?

TIDAL PREDICTIONS

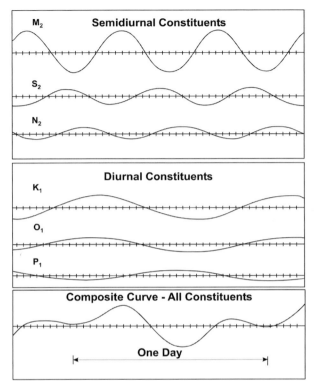

Figure 6.6 Six tidal constituents and the resulting wave form (Gill and Schultz, 2001).

The answer lies in topography, specifically the length and depth of the ocean basins, the geometry of the coastline and the width and depth of the continental shelf. Most readers will have watched as wind-generated waves approach a beach from the ocean and slowly change their shape and size (shoaling) and direction (refraction and reflection) as they enter shallower water. In the same way, the tidal wave is greatly affected by shallow water and the geometry of the coastline. But as in all discussions of waves, the definition of "shallow" is relative, and is largely dependent on the length of the wave. Since the tidal wave is such a long wave, with a period (time between crests) of 12–24 hours as opposed to 24 seconds for even the longest wind waves, the tidal wave does not shoal or refract near the beach; it actually begins to transform as it approaches a continental shelf! As a result, the features of the tide vary dramatically along coastlines around the world, depending on the ocean basin, the orientation and geometry of the coastline, and the bathymetry of the coastal ocean out to the continental

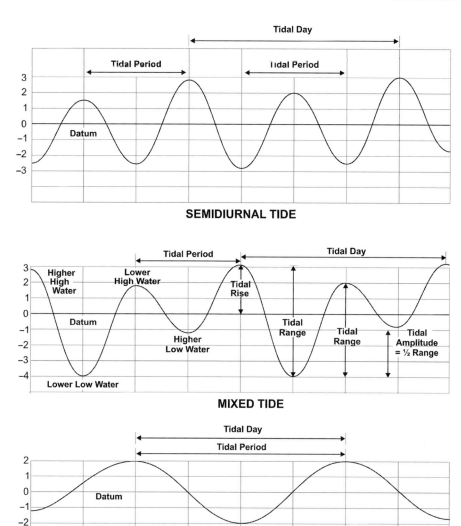

Figure 6.7 Three dominant tide types (Gill and Schultz, 2001).

shelf break. Particularly high tide ranges can be produced in enclosed basins (such as the Bay of Fundy in Nova Scotia) as a result of water being funneled into a narrow, shallow embayment and/or oscillations (hence, amplification) triggered by the tidal wave when the tidal wave is in resonance with the enclosed body of water.

The science of tidal prediction goes back to the observations of Galileo and Kepler, and the mathematical descriptions developed by Newton. Many others

Table 6.1 *Mean tidal range at various locations around the Earth*

Station	Mean range (feet)
Burntcoat Head, Minas Basin, Bay of Fundy, Nova Scotia	38.4
Port of Bristol (Avonmouth), United Kingdom	31.5
Sunrise, Turnagain Arm, Cook Inlet, Alaska	30.3
Rio Gallegos (Reduccion Beacon), Argentina	29.0
Granville, France	28.2
Banco Direccion, Magellan Strait, Chile	28.0
Cape Astronomicheski, Kamchatka, Russia	24.1

From NOAA (2018d).

followed, including Laplace, Kelvin, and Doodson, eventually resulting in the development of predictive algorithms for the tide elevations at locations around the Earth, including the coastal topographic influences discussed here. As a result, tidal predictions are available in published tide tables and websites (e.g., https://tidesandcurrents.noaa.gov/tide_predictions.html in the United States and www.ntslf.org/tides/predictions in the UK).

6.2.1   Tidal Datums

Tidal Datums are employed as reference points from which to measure water depths and elevations, e.g., for navigation charts that guide mariners in understanding the minimum expected water depth at a particular location. The Datums are established by measuring the tidal record over an extended period of time, known as the Tidal Datum Epoch. The US National Oceanic and Atmospheric Administration (NOAA) employs a 19-year National Tidal Datum Epoch, a time period that approximates the variation in the path of the moon about the sun. We list here the various Tidal Datums employed by NOAA's Center for Operational Oceanographic Products and Services (https://tidesandcurrents.noaa.gov/datum_options.html):

> Highest astronomical tide: The elevation of the highest astronomical tide expected to occur at a specific tide station over the National Tidal Datum Epoch.
>
> Mean Higher High Water (MHHW): The average of the higher high water height of each tidal day observed over the National Tidal Datum Epoch. For stations with shorter series, comparison of simultaneous observations with a control tide station is made in order to derive the equivalent Datum of the National Tidal Datum Epoch.

Mean High Water (MHW): The average of all the high water heights observed over the National Tidal Datum Epoch. For stations with shorter series, comparison of simultaneous observations with a control tide station is made in order to derive the equivalent Datum of the National Tidal Datum Epoch.

Mean Tide Level (MTL): The arithmetic mean of mean high water and mean low water.

Mean Sea Level (MSL): The arithmetic mean of hourly heights observed over the National Tidal Datum Epoch. Shorter series are specified in the name; e.g., monthly mean sea level and yearly mean sea level.

Mean Low Water (MLW): The average of all the low water heights observed over the National Tidal Datum Epoch. For stations with shorter series, comparison of simultaneous observations with a control tide station is made in order to derive the equivalent Datum of the National Tidal Datum Epoch.

Mean Lower Low Water (MLLW): The average of the lower low water height of each tidal day observed over the National Tidal Datum Epoch. For stations with shorter series, comparison of simultaneous observations with a control tide station is made in order to derive the equivalent Datum of the National Tidal Datum Epoch.

Lowest Astronomical Tide (LAT): The elevation of the lowest astronomical tide expected to occur at a specific tide station over the National Tidal Datum Epoch.

The various Datums are illustrated in Figure 6.8.

6.3   Atmospheric and Wind Effects on Water Level

Returning to Figure 6.5, it is clear that there are influences at work other than the complex combination of periodic constituents associated with the tide. There appear to be segments of the record where there is a superimposed rise or fall of the water level over short periods of time (order of days). The most likely contributor here, and in many other water level records, is the influence of the wind and atmospheric pressure.

As has already been discussed, surface currents are generated by the action of the wind over the water. These winds are associated with pressure fields in the atmosphere. Figure 6.9 illustrates an idealized example of a gradient in the atmospheric pressure. Since P_2 is greater than P_1, we expect that the flow of air

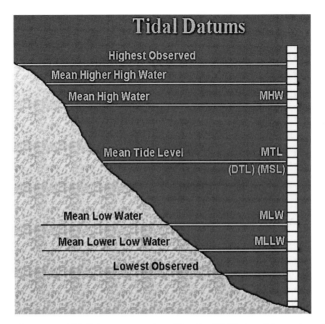

Figure 6.8 Tidal Datums employed by the US National Oceanic and Atmospheric Administration (NOAA, 2014).

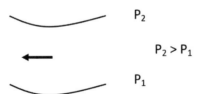

Figure 6.9 Wind direction associated with atmospheric pressure gradient (Northern Hemisphere).

(the wind) will be directed from 2 toward 1, much as water flows downhill. However, at the scale of atmospheric motion, the Coriolis "force" plays a dominant role, and in the Northern Hemisphere the flow is directed to the right of the high-pressure area, as depicted here. In the Southern Hemisphere, the flow direction is reversed.

This pattern is observed most dramatically in cyclones, which are atmospheric systems characterized by a low pressure at the center, and anticyclones, characterized by a high pressure at the center. Figure 6.10 illustrates these phenomena, depicted here for the Northern Hemisphere. Again, the flow is reversed in the Southern Hemisphere.

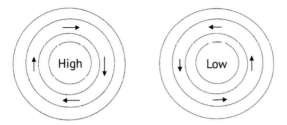

Figure 6.10 Wind direction associated with anticyclone (left) and cyclone (right), in the Northern Hemisphere.

These atmospheric pressure features can produce sometimes dramatic changes in the water level along the coast. The water level changes are caused by many different mechanisms, some obvious and others not so obvious. First the obvious: onshore and offshore winds directly influence coastal water levels by either pushing water toward the coast and causing an increase in water levels, in the case of onshore winds, or by pushing water away from the coast and causing a decrease in water levels, in the case of offshore winds. Atmospheric pressure can play a role as well. Consider the cyclone and anticyclone systems mentioned earlier. Across the scale of these systems, high pressure can cause a measurable decrease in the water level while low pressure can cause a measurable increase in the water level. For atmospheric systems with extremely low pressure, the increase in the water level can be significant. Table 6.2 indicates the water level increase associated with various central pressures of cyclones (US Army Corps of Engineers, 1989).

An additional influence of the wind that is discussed elsewhere in this book is upwelling and downwelling. We know that large-scale persistent winds along the coast produce vertical currents and either offshore (downwelling) or onshore (upwelling) water flows. This phenomenon also causes a large-scale change in the water level along the coast, with downwelling being associated with a rise in the water level and upwelling a decrease in the water level.

6.3.1   Storm Surges

In the context of urban coastal regions, we must consider several other, less persistent, less predictable, but potentially far more impactful water level phenomena. These include storm surges caused by tropical and extratropical cyclones. As has already been explained, a cyclone can produce a significant increase in water level because of the reduced pressure at the center. When this low pressure center moves, as is the case for a hurricane (defined as a cyclone with winds exceeding 75 miles per hour), the elevated water level moves with the storm. In the case of a powerful cyclone, this atmospheric pressure effect can

Table 6.2 *Water level increase associated with atmospheric pressure, relative to water level at atmospheric pressure of 1013 mb, or 29.91 inches of Hg*

Central pressure of storm (mb)	Central pressure of storm (inches of Hg)	Water level increase (feet)
900	26.58	3.78
910	26.87	3.45
920	27.17	3.11
930	27.46	2.78
940	27.76	2.44
950	28.05	2.11
960	28.35	1.77
970	28.64	1.44
980	28.94	1.10
990	29.23	0.77
1000	29.53	0.43

From US Army Corps of Engineers (1989).

cause an increase in the sea surface elevation of several feet, across an area on the order of 50 – 100 miles. As the cyclone approaches shore, this elevated sea level behaves like a wave, and just as we learned in the case of the tide, the wave begins to "feel" the bottom and shoal, with an increase in amplitude that depends on the speed and direction of the storm.

Added to this traveling bulge of water is the water level increase associated with onshore winds along that area of the cyclone where the winds are directed toward the shore. Again, depending on the speed and direction of the storm, this component of the cyclone's influence can cause an increase in sea surface elevation of several feet. Of course, we would also expect a strong cyclone to produce high waves. These waves also produce a large-scale persistent increase in the water level at the coast, termed wave setup, with elevation increases of a foot or more not uncommon. Taken together, these water level increases combine to produce the phenomenon known as a storm surge. These sometimes catastrophic events occur around the world, virtually anywhere an extreme windstorm occurs but most commonly wherever strong cyclones occur. The increase in water level, or "surge", can vary from a few feet to several meters, and can last for a few hours to a day or more, depending on the speed and direction of the storm.

As an illustration of the influence of a storm surge on water levels along an urban coastline, we examine the impact of Hurricane Sandy on the coastline of New York City in October, 2012. Figure 6.11 illustrates the track of the storm from October 22 to October 31. Note that just before landfall, the atmospheric

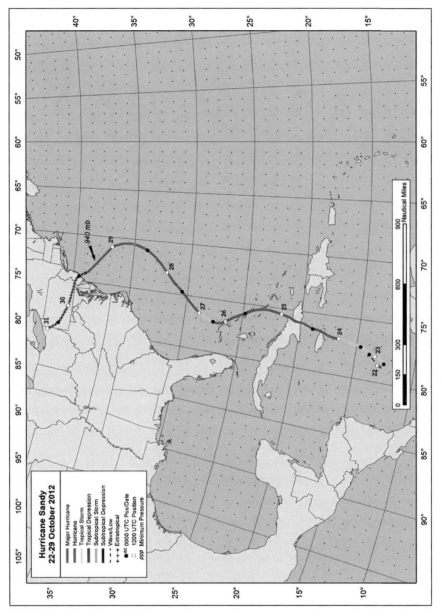

Figure 6.11 Hurricane Sandy track, from October 22 to October 31, 2012, with date indicated along the track (Blake et al., 2013).

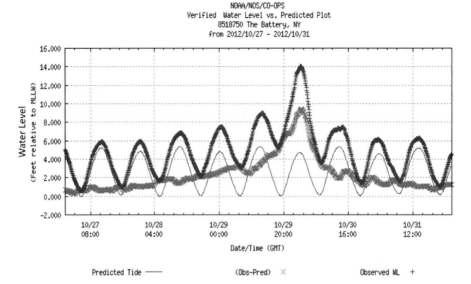

Figure 6.12 Water elevation at The Battery in lower Manhattan during Hurricane Sandy, October, 2012, with the blue line representing the predicted tide, the red line representing the measured water elevation, and the green line representing the difference between the expected (predicted) water elevation and the measured water elevation (NOAA, 2012). (A black-and-white version of this figure appears in some formats. For the color version, please refer to the plate section.)

pressure at the storm center was 940 mb, which according to Table 6.2 would have caused a rise in the sea level of nearly 2.5 feet alone. The path of the storm placed the New York metropolitan area right in the path of the most damaging zone of the hurricane – the northeast quadrant, where onshore winds and the forward movement of the storm combined to produce the highest winds.

We illustrate in Figure 6.12 the water elevation record at The Battery, located at the southern tip of Manhattan, during Hurricane Sandy. Here, the blue line represents the predicted water level associated with the tide. The red line is the actual, recorded water level, and the green line depicts the difference between the actual water level and the predicted tidal water level elevations. Hence, the green line represents the measured storm surge. The peak storm surge measured approximately 9.5 feet and happened to occur at almost exactly the time of high tide. The combined water level elevation of just over 14 feet relative to MLLW was the highest water level ever recorded in lower Manhattan. Note also that the influence of the hurricane on the water level was observed more than 2 days prior to the arrival of the storm, as the storm traveled northward along the East Coast. We will discuss the impact of Hurricane Sandy in Chapter 11.

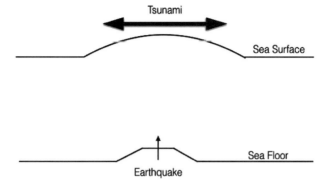

Figure 6.13 Depiction of the generation of a tsunami by an underwater earthquake.

6.4   Tsunamis

A tsunami (Japanese for "harbor wave") is a wave caused by an underwater disturbance, including earthquakes, submarine landslides, and volcanic eruptions, although earthquakes are the most common generating mechanism. Large tsunamis are generally caused by the "dip-slip" fault type of earthquake, where there is a sudden upward movement of the seafloor on one side of the fault and a downward movement of the seafloor on the other side of the fault. As depicted in Figure 6.13, which could be interpreted as the upward motion portion of a dip-slip fault type of earthquake (or perhaps the sudden placement of material at the bottom due to a landslide), the underwater disturbance in essence displaces a volume of water at the surface. It is reasonable to treat the volume of the displaced water at the surface as being equal to the volume of the uplifted seafloor bottom. This displaced volume then propagates outward in all directions as a series of stable waveforms, commonly referred to as "solitons". The characteristics of the tsunami wave train are a complex function of the generating mechanism (e.g., the underground depth, magnitude, and type of earthquake), the area and height of the displaced seafloor, the water depth at the location of the earthquake, and of course the bathymetry and geometry of the coastal region in the vicinity of the earthquake.

The nature of the generation mechanism causes this type of wave to be much longer than wind-generated waves, with wavelengths on the order of 100 miles. Depending on the size of the underwater disturbance, the height of the wave can vary from a few inches to several feet. This combination of very long length and relatively low height makes the tsunami a very difficult wave to detect in the open ocean. However, as the tsunami approaches a coastline, it begins to feel the bottom at great depths, and gradually begins to shoal. The exact form and impact of a tsunami on a particular coastline is difficult to predict without sophisticated computer models, but we can expect that these long waves experience significant transformation as a result of interaction with the continental shelf, nearshore reefs should they exist, as well as headlands and/or bays. Wave reflection, especially along shorelines with steep slopes, is possible. The wave height of a tsunami at the coast is often measured in 10s of feet, with wave run-up (maximum elevation of the wave onrush after breaking) as high as 100 feet or greater, as occurred in the Aleutian Islands earthquake and tsunami on April 1, 1946, which caused wave run-ups along the coast reaching nearly 140 feet. More recently, the devastating Indian Ocean tsunami of December 26, 2004 produced wave run-ups of more than 130 feet along several coastlines and killed at least 225,000 people in 15 countries. The March 11, 2011, Tōhoku, Japan earthquake was the largest magnitude earthquake in the history of Japan and generated a tsunami that had maximum run-up heights of just under 130 feet, killing more than 18,000 people and causing more than $220 billion in damages, making it the single most expensive natural disaster in history (see http:// itic.ioc-unesco.org).

Although much attention is properly placed on the wave height and run-up along a coast when the tsunami wave crest arrives at the shoreline, one must also consider the impact of the arrival of the tsunami wave trough. Depending on the generation mechanism, the tsunami may arrive crest-first or trough-first. But in any case, the wave trough will arrive at some point during a tsunami event, and the extremely low water level at the shoreline can itself produce severe adverse impacts. These impacts include damage to coastal structures such as seawalls and breakwaters due to the sudden removal of water along the structure, resulting in a dramatic and nearly instantaneous force imbalance, as well as possible impact to the underlying soil and hence the stability of the structure foundation. Should the tsunami occur in the vicinity of a harbor, vessels can be left stranded on the dry seafloor, a condition for which ships are not designed. Very high currents can arise during the offshore rush of water as the wave trough arrives, producing beach erosion and scouring, including along submerged pipelines and other infrastructure, which can experience sudden failure as a result.

Although tsunamis can occur anywhere in the world, the great majority occur in the Pacific because of the much higher level of tectonic activity in the Pacific. Figure 6.14 illustrates the locations of the measured major earthquakes (magnitude 7.0 or higher) around the world between July 29, 1900 and November 12, 2017 (https://earthquake.usgs.gov/earthquakes/browse/). Note the very high concentration of earthquakes along the so-called "Ring of Fire" in the Pacific Ocean.

Because of the significant destructive potential and threat to life that tsunamis represent, there has long been international cooperation in providing tsunami

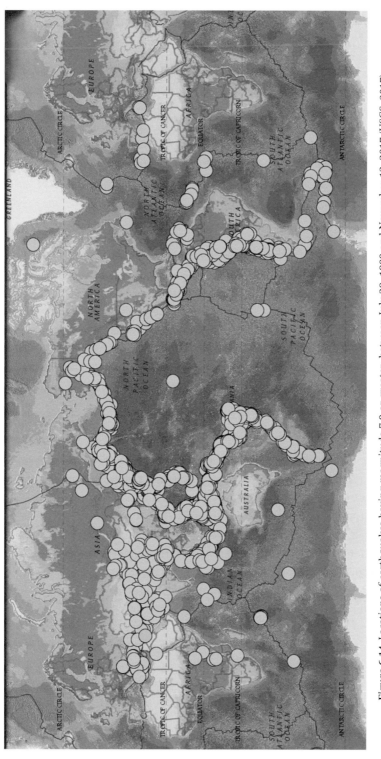

Figure 6.14 Locations of earthquakes having magnitude 7.0 or greater, between July 29, 1900 and November 12, 2017 (USGS, 2017).

warnings to coastal communities around the world. The International Tsunami Information Center is located in Honolulu, Hawai'i. The Intergovernmental Coordination Group for the Pacific Tsunami Warning and Mitigation System (ICG/PTWS) is an international effort involving many nations in the Pacific, with the mission to coordinate activities and improve tsunami warning and preparation capabilities. The participants in the ICG/PTWS include the United States, Australia, Brunei Darussalam, China, Columbia, Costa Rica, France, Guatemala, Honduras, Japan, Kiribati, Mexico, Nicaragua, Palau, Panama, Peru, Russia, Samoa, and Singapore.

6.5   Emerging Opportunities

We have discussed much about the risks associated with water level changes, in particular the water level changes associated with extreme events such as storm surges and tsunamis. But the rise and fall of the ocean surface also provides many opportunities for benefit to the urban populations that line the coast. Perhaps chief among these benefits is the potential for renewable energy from tidal water elevation changes and/or the associated currents. Since the astronomical tides are well understood and predictable as discussed earlier, the potential for viable (profitable) investment in tidal power generation can be assessed in a relatively straightforward manner.

There are numerous examples of tidal power generation around the world. Many if not most of these facilities are located in regions where either the tide range or the tide-generated currents are sufficient to warrant commercial investment. Of course, other considerations can dominate the decision-making process, including regulatory requirements, the investment environment, etc.

The Sihwa Lake Tidal Power Station in South Korea, completed in 2011, has the largest electricity generation capacity at 254 megawatts (MW). The facility makes use of a pre-existing dam structure and 10 turbines that capitalize on the flow associated with an 18-feet tide range. The oldest operating tidal power plant is in La Rance, France, with 240 MW of electricity generation capacity.

Tidal power is still in its infancy, for many reasons. First and perhaps foremost is the requirement that the location of a commercially viable tidal power source (i.e., a location with high tide range and/or strong tidal currents) is near a population center or industrial facility with powering needs sufficient to warrant the significant financial investment associated with these facilities. Another challenge is the potential for adverse environmental impacts, for example from the construction of a dam across the tidal river or estuary. This structure, often referred to as a "barrage", allows the water to flow upstream through open gates as the tide rises. The gates are then closed at the time of high tide, and a large volume of water is trapped in a pool or lake behind the barrage. The water is then released downstream through turbines. This design is quite efficient, but it has adverse environmental impacts because of abrupt changes in the land and water area behind the structure, which can severely impact the native land and aquatic life. The system can also lead to rapid changes in the salinity and turbidity of the water on both sides of the barrage. And, of course, there is always the concern about marine animals being caught in the turbine blades, although this impact can be mitigated by screens and other techniques. An alternative approach, that of using free-standing turbines in a tidal river or estuary, akin to a wind turbine on land, has found success in several applications around the world. This approach appears to have minimal adverse environmental impacts, as the movement of the turbine blades with the water motion is slow enough that marine animals can avoid impact.

As coastal urban population centers continue to grow, so too will their need for sustainable and continuous power. In those areas where the tide range and/or tidal currents can produce commercially viable tidal power, cities will undoubtedly turn to the ocean as an essential component of their future energy strategy.

7

Estuarine and Coastal Ocean Flows

7.1 Introduction

The coastal ocean and its weather patterns drive processes and events that range from highly supportive of human populations (e.g., via fishing, marine transportation, and tourism) to highly threatening (e.g., via storm surges and flooding). Those processes lead to a rich combination of current patterns and distributions of temperature and salinity. The equations governing the motion in the urban ocean, the estuaries and coastal ocean, are very complex. They are nonlinear and coupled. No closed form solutions exist, except in the simplest of cases. Consider a typical urban ocean where we let H represent the approximate water depth and L the approximate horizontal scale of the area of interest. The depth H is typically $H \sim 10$ m and $L \sim 10$ km. The ratio, $H/L \sim 10^{-3}$, which is a very small number; i.e., the coastal ocean is very shallow. In view of the complexity of the equations of motion, it is worthwhile to inquire into the possible simplification of the equations based on this geometrical fact.

7.2 Estuarine Circulation

It is in the estuaries where man has his most intimate contact with the marine environment. An estuary is a narrow, semienclosed coastal body of water that has a free connection with the open sea at least intermittently and within which the salinity of the water is measurably different from the salinity in the open ocean (Pritchard, 1967). The estuary has fresh water overriding salty ocean water moving in the opposite direction. Where they first meet is called the limit of saltwater intrusion. The presence of tidal flow generates turbulence through

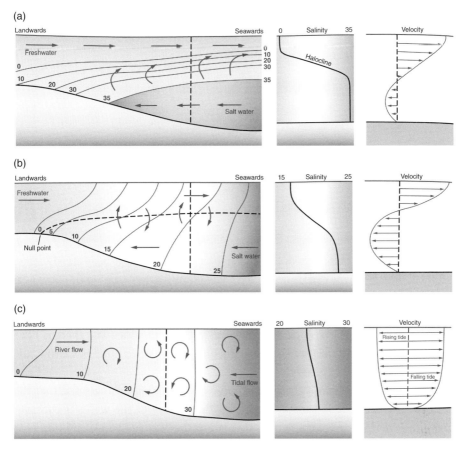

Figure 7.1 Schematic of the physical processes in a partially mixed estuary. Mean flows in (a) and (b) are seaward at the surface and landward at the bottom.

the entire water column (the turbulence is induced mainly at the bottom). As a result, salt water from the ocean is stirred into the upper layer and fresh water into the lower layer. Salinity therefore changes along the estuary axis not only in the upper layer but also in both layers. Figure 7.1 illustrates the estuarine sequence (Pritchard 1955). Increasing tidal mixing readily weakens the halocline and creates a whole sequence of estuarine behavior from a salt wedge estuary (e.g., Mississippi River), to a moderately stratified estuary (e.g., Hudson River, Chesapeake Bay), to a vertically homogeneous estuary (e.g., Delaware Bay, Tampa Bay). The relations developed by Knudsen in Chapter 5 work best for these salt wedge estuaries. It should be noted that as the halocline weakens, there is less salinity intrusion and typically larger flows in the upper and lower layers.

The two-layer vertical structure of estuarine flows can be readily explained by simplifications to the governing equations. The equations governing the estuary as derived in the previous two chapters are

$$\frac{Du}{Dt} = -\frac{1}{\rho_0}\frac{\partial p}{\partial x} + fv + A_H \nabla_H^2 u + \frac{\partial}{\partial z}K_M\frac{\partial u}{\partial z} \tag{7.1}$$

$$\frac{Dv}{Dt} = -\frac{1}{\rho_0}\frac{\partial p}{\partial y} - fu + A_H \nabla_H^2 v + \frac{\partial}{\partial z}K_M\frac{\partial v}{\partial z} \tag{7.2}$$

$$0 = -\frac{1}{\rho}\frac{\partial p}{\partial z} - g, \tag{7.3}$$

$$\frac{\partial u}{\partial x} + \frac{\partial v}{\partial y} + \frac{\partial w}{\partial z} = 0 \tag{7.4}$$

Let us assume the velocities are not changing in time, and it is at steady state. All the tendency terms can be neglected. Now let us further assume the flow is in a narrow body of water of width B. The narrowness means that the lateral gradients, d/dy, are much smaller than those along the channel gradients. U, for example, is about the same all the way across that channel, and there is little cross channel velocity, v, that is, $v = 0$. And given that the channel is narrow, the Coriolis force can be neglected too.

The horizontal pressure gradient can also be simplified by assuming that the atmospheric pressure is constant and that the density of the water is constant along the channel and with depth. This is a rather restrictive assumption that will be relaxed later. The barometric and the baroclinic pressure gradients can then be neglected as

$$-\frac{1}{\rho}\frac{\partial P}{\partial x} = -\frac{1}{\rho}\left[\frac{\partial P_a}{\partial x} + \rho g\frac{\partial \eta}{\partial x} + g\int_z^0 \frac{\partial \rho}{\partial x}dz\right] \tag{7.5}$$

The governing equations then end up with a balance between the horizontal pressure gradient – the slope of the free surface – and the vertical friction (mixing) terms:

$$g\frac{\partial \eta}{\partial x} = \frac{\partial}{\partial z}k_m\frac{\partial u}{\partial z}. \tag{7.6}$$

Assume that the vertical mixing, K_M, is constant with depth. This is a weak assumption because when the tides are active or the wind is blowing, we have a lot more mixing at the surface or a lot more mixing at the bottom. But for now, we just say it is constant everywhere. Integrating twice with respect to z, we obtain

$$u(z) = \frac{g}{2k_m}\frac{\partial \eta}{\partial x}z^2 - C_1 z + C_2, \tag{7.7}$$

where the constants of integration C_1 and C_2 are determined from surface and bottom boundary conditions. At the surface, we assume that there is a wind inducing a frictional stress as

$$k_m \frac{\partial u}{\partial z} \bigg|_{surf} = \frac{\tau_W^x}{\rho_0}, \tag{7.8}$$

and at the bottom the velocity is 0, $u(-H) = 0$. The wind we know from data. The problem is knowing what the slope of the water surface is. We are going to get that by looking at conservation of mass. When we go out to an estuary like the Hudson River, we know there is a river coming down with fresh water. Call the amount of freshwater Q (m³/s). This fresh water coming down tends to push the saltier water out of the estuary, but the ocean is constantly pushing the salt back in. This tug of war is the essence of estuarine circulation. Solving for the two constants of integration using the boundary conditions, we obtain

$$u(z) = \frac{g}{2k_m} \frac{\partial \eta}{\partial x} \left(z^2 - H^2 \right) - \frac{\tau_W^x}{k_m \rho_0} (z + H). \tag{7.9}$$

Conservation of volume in steady state, as we learned in Chapter 5, states that the volume of water entering the estuary equals the volume of water leaving it. The water entering the bottom layer will eventually work its way into the top layer for its exit to the ocean. Integrating the velocity in Equation 7.9 from surface to bottom and then across the estuary, the result will be the river flow Q.

This provides a relationship for the slope of the free surface:

$$g \frac{\partial \eta}{\partial x} = -\frac{3q}{H^3} - \frac{3\tau_W^x}{2k_m \rho_0} \frac{1}{H}, \tag{7.10}$$

where $q = Q/B$. Now, we can replace the surface slope in Equation 7.9 by the relationship shown in Equation 7.10 yielding

$$u(z) = \frac{3}{2} u_0 \left(1 - \sigma^2 \right) + \frac{1}{4} \frac{\tau_W^x H}{k_m \rho_0} \left(1 + 4\sigma + 3\sigma^2 \right) \tag{7.11}$$

$$\left[\sigma = \frac{z}{H} \right] \text{ goes from } -1 \rightarrow 0$$

where u_0 is the surface current due to the river flow. The total velocity is proportional to the river flow and the wind stress. The vertical distribution of the river flow alone (assuming no wind) is shown in Figure 7.2, and the effect of a wind blowing downstream is shown in Figure 7.3. The effect of a wind blowing upstream is similar to the distribution shown in Figure 7.3, except that the currents are reversed.

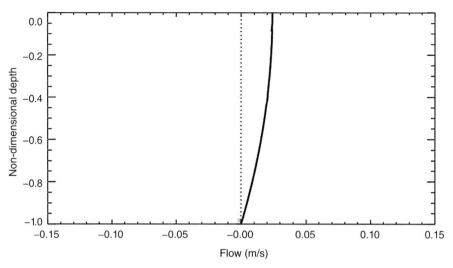

Figure 7.2 Velocity distribution of the river flow in an idealized estuary. A vertical integration yields the total river flow Q.

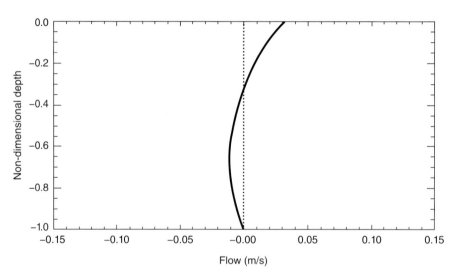

Figure 7.3 Velocity distribution of the river flow in an idealized estuary with a wind blowing in the downstream direction. The vertical integration of the velocity yields no net flow.

The vertical distribution of the wind-produced currents is controlled by the vertical mixing, the turbulence, which appears in the denominator of Equation 7.10. This says that when the level of turbulence mixing is large, the velocity becomes smaller as it becomes spread out more uniformly in the vertical.

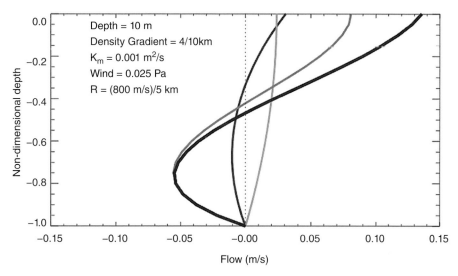

Figure 7.4 The vertical distribution of all the components of the estuarine circulation. The density gradient contribution is in orange; the wind is in red; the river flow is in green. The net flow is in blue, surface outflow and bottom inflow. (A black-and-white version of this figure appears in some formats. For the color version, please refer to the plate section.)

Now, let's relax the assumption where the density is constant and allow for horizontal changes in density where the density is largest near the ocean and decreases up the estuary. This gradient is the same from surface to bottom. The governing equation now becomes

$$0 = \frac{g}{\rho_0} \int \frac{\partial p}{\partial x} dz + g \frac{\partial \eta}{\partial x} + k_m \frac{\partial^2 u}{\partial z^2}, \tag{7.12}$$

whose solution becomes after applying the surface and bottom boundary conditions from above,

$$U(Z) = \frac{3q}{2}\left(1 - \sigma^2\right) + \frac{1}{48} \frac{gH^3}{\rho_0 K_m} \frac{\partial \rho}{\partial x}\left(1 - 9\sigma^2 - 8\sigma^3\right) + \frac{1}{4} \frac{\tau_W^x H}{k_m \rho_0}\left(1 + 4\sigma + 3\sigma^2\right). \tag{7.13}$$

Once again, the vertical mixing appears in the denominator. This solution is plotted in Figure 7.4 in an estuary of 10 m depth, a downstream blowing wind, and a level of turbulence characterized by $K_M = 1 \times 10^{-3}$ m²/s, a typical value.

The estuarine dynamics are now clear. There is a distinct vertical structure to the flow. There is flow in the upper layer toward the ocean (saltier end) and flow in the bottom toward the fresher end, the river. The magnitude of the currents is controlled by the turbulence mixing. When it is large, the velocity becomes smaller as it becomes spread out more uniformly in the vertical.

Estuarine variability occurs on both short time scales of 12–25 hours and longer time scales of greater than say 10 days. The 10-day time scales are due to wind events both local and remote as shown in Figure 2.2. Water coming in from the sea is much saltier than that in the estuary, greatly affecting the vertical structure of the salinity and the currents. Also in this category are variations driven by weather systems. At the longer time scales, the most dramatic change in an estuary is the rapid response to changes in river inflow related to severe storms and rapid snow melt events. For example, the response of the salinity in Chesapeake Bay to changes in river flow is shown in Figure 7.5. The longitudinal variation in surface salinity over the length of the bay ranges from almost 0 ppt near the upper end to a salinity of about 25–30 ppt at the ocean end. The longitudinal distribution in the bay for a high flow period is shown in the top panel of Figure 7.5 and for a period of more normal river flow in the bottom panel of Figure 7.5. Not only is the halocline sharpened up during high river flow but the stratification is greatly enhanced.

The flow in an estuary is confined because lateral boundaries are important to how we think about their dynamics. The water primarily flows horizontally and vertically along the estuary. Now, let us move out of the estuary onto the continental shelf to examine the circulation in that part of the urban ocean. The shelf areas typically range in depths from ten to a few hundred meters before encountering the shelf break where the depth plunges quickly to 1,000 m. Generally, they extend 10–100 km away from the coast. The coastlines themselves can be quite complicated with capes, promontories, and embayments; continental shelf bottoms can be flat or complex, with features such as canyons, sea mounts, and ridges. On the wide open shelf region, the water motion is three-dimensional.

7.3   Continental Shelf Circulation

The circulation over the Middle Atlantic Bight off the coasts of New York and New Jersey is shown in Figure 7.6. It is illustrated using satellite tracked drifters. The mean circulation shows a persistent alongshore (southwestward) system of currents even though the mean wind is eastward. An alongshore pressure gradient, with mean sea level decreasing by about 10 cm between Cape Cod and Cape Hatteras (i.e., along the flow direction), is the reason. The source of this alongshore pressure gradient is not the coast circulation but rather the gyre circulations of the North Atlantic Ocean. Superimposed on this mean current are the presence of loops that the drifters are exhibiting. These inertial oscillations and inertial currents are very commonly observed in the offshore waters of the urban ocean.

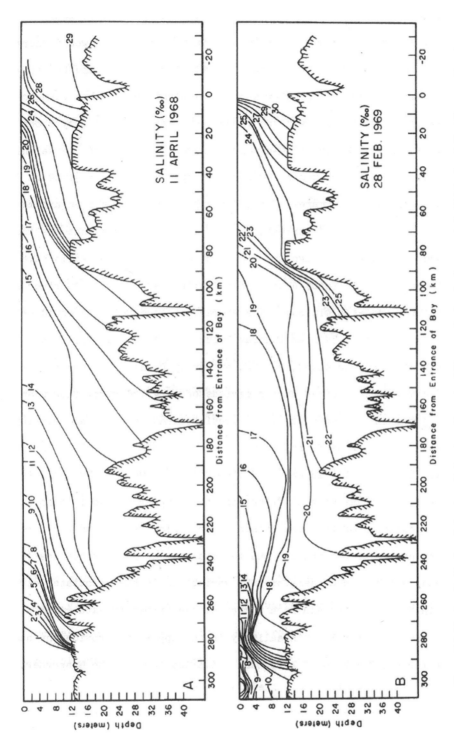

Figure 7.5 The longitudinal variation in salinity along the axis of Chesapeake Bay during a period of high river inflow (top panel) and more normal flow (bottom panel) (Seitz, 1971).

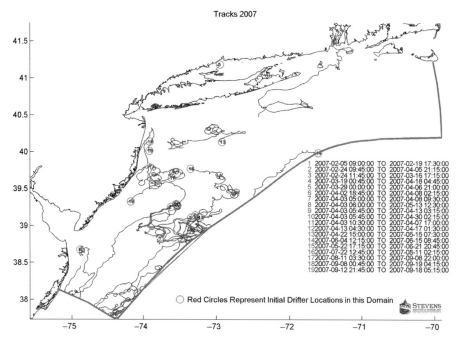

Figure 7.6 Currents deduced from satellite tracked drifters off the coasts of New York and New Jersey. The red line is the shelf break where water depths drop rapidly from 100 m or so to 1,000 m. (A black-and-white version of this figure appears in some formats. For the color version, please refer to the plate section.)

Why are the drifters undergoing these oscillations? To figure that out let us go back to our equations. Out on the continental shelf break, the gradient of pressure horizontally is the same (ignoring for now the prevailing alongshore gradient) and is small. The wind is often weak so that the vertical mixing is small. The equations are then reduced to

$$\frac{Du}{Dt} = -\frac{1}{\rho_0}\frac{\partial p}{\partial x} + fv + A_H \nabla_H^2 u + \frac{\partial}{\partial z}K_M\frac{\partial u}{\partial z} \tag{7.14}$$

$$\frac{Dv}{Dt} = -\frac{1}{\rho_0}\frac{\partial p}{\partial y} - fu + A_H \nabla_H^2 v + \frac{\partial}{\partial z}K_M\frac{\partial v}{\partial z} \tag{7.15}$$

and by neglecting the advection terms we obtain,

$$\frac{\partial u}{\partial t} - fv = 0, \tag{7.16}$$

$$\frac{\partial v}{\partial t} + fu = 0. \tag{7.17}$$

To solve those equations, we take the time derivative of Equation 7.16 and 7.17 and rearranging we obtain

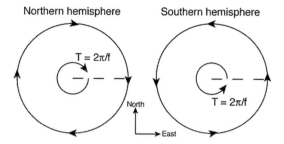

Figure 7.7 Particle paths undergoing inertial oscillations in the Northern Hemisphere (left side) and in the Southern Hemisphere (right side).

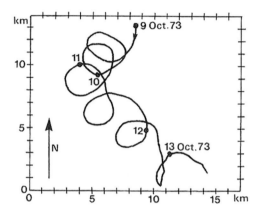

Figure 7.8 Drifter path in the Mediterranean Sea at Latitude 43.15° N.

$$\frac{\partial^2 u}{\partial t^2} + f^2 u = 0 \tag{7.18}$$

and similarly,

$$\frac{\partial^2 v}{\partial t^2} + f^2 v = 0, \tag{7.19}$$

with solutions are $u = U_o \sin(ft)$ and $v = U_o \cos(ft)$, where U_o is an arbitrary amplitude. Together these velocities will move a water parcel in a circle called inertial oscillations, of radius $r = U_o/f$ and with a period $T = 2\pi/f$, as shown in Figure 7.7. In the Northern Hemisphere, the inertial oscillations move particles in a clockwise path, while in the Southern Hemisphere the opposite occurs.

Since the motion on the continental shelf is so complex, the question arises whether the particle paths are due to inertial processes instead of winds or tides or unusual currents. Consider the particle paths of a drifter in the Mediterranean Sea at Latitude 43.15° N, as shown in Figure 7.8. Counting the day marks along the curve shows the number of loops that a particle makes and in a certain time, i.e., 4 days. There were six loops resulting in 16 hours per loop. Now compute the inertial period and compare those times. If times are close then the loops are inertial. At a latitude of $43.15°$ N, $f = 1. \times 10^{-4}\,s^{-1}$ so that the interval period $T = 2\,pi/f = 18$ hours, which matches the period determined by the observations. We can also estimate the velocity of the drifters as they undergo the oscillations from the radius of the loops in Figure 7.8, which is 1.5 km, and then work backwards to get 0.15 m/s.

7.4   Coastal Upwelling and Downwelling

Wind-driven currents often tend to dominate the flow patterns that are commonly observed in the coastal ocean. Ekman (1905) performed a remarkable series of studies during the first half of the twentieth century using a simplified set of equations that led to an understanding of how winds drive the urban ocean's circulation, and to the concepts of upwelling and downwelling. He assumed a balance between the Coriolis force, the vertical mixing, and the frictional force, as in

$$\frac{Du}{Dt} = -\frac{1}{\rho_0}\frac{\partial p}{\partial x} + fv + A_H \nabla_H^2 u + \frac{\partial}{\partial z} K_M \frac{\partial u}{\partial z}, \tag{7.20}$$

$$\frac{Dv}{Dt} = -\frac{1}{\rho_0}\frac{\partial p}{\partial y} - fu + A_H \nabla_H^2 v + \frac{\partial}{\partial z} K_M \frac{\partial v}{\partial z}, \tag{7.21}$$

$$0 = -\frac{1}{\rho}\frac{\partial p}{\partial z} - g, \tag{7.22}$$

$$\frac{\partial u}{\partial x} + \frac{\partial v}{\partial y} + \frac{\partial w}{\partial z} = 0, \tag{7.23}$$

where the boundary conditions are that the frictional forces at the surface are due to the wind. According to Ekman's theory, which is too lengthy and detailed to provide here (see Cushman-Roisin and Beckers, 2011 for a thorough review), the effect of the wind is to drive water to the right of the wind in the Northern Hemisphere. Coastal upwelling occurs when a wind blows parallel to the shore, as shown in Figure 7.9a. The transport (vertical integration of the velocity) in the surface layer is to the right of the wind causing deeper water that is colder than the surface waters to upwell into the near-surface. Coastal downwelling occurs when the wind blows in the opposite direction, as in Figure 7.9b. The transport in the surface layer is again to the right of the wind but now toward the shore, bringing typically warmer offshore waters to the coast. Here the velocity near the coast is downward carrying the warmer water with it. Winds that blow toward the shore and away from it play little role in creating currents close to the near coast.

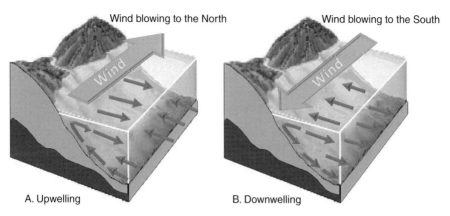

Wind blowing to the North   Wind blowing to the South

A. Upwelling   B. Downwelling

Figure 7.9 Coastal upwelling in the northern hemisphere with (a) wind blowing to the north and coastal downwelling with (b) wind blowing to the south. In the southern hemisphere, the relationship between the wind and currents is opposite to that shown (courtesy of Victor Rodriguez and Firas Saleh). (A black-and-white version of this figure appears in some formats. For the color version, please refer to the plate section.)

7.5   Coastal Plumes and Fronts

Most prominent along the urban oceans of the globe are river discharges, which can lead to the dramatic features observed in the coastal ocean. The discharges with their fresher than coastal ocean waters produce strong coastal currents and create intense fronts. Fronts, defined as regions where water properties change markedly over a relatively short distance, may only be a few meters wide. What these regions have in common is the change of some oceanographic properties – temperature, salinity, etc. – across the width of the front is an order of magnitude larger than changes of the same property over a similar distance on either side of or along the front. An understanding of frontal dynamics is important for managing how urban centers interact with the coastal ocean. Floating detritus and particulate matter tends to accumulate in fronts, and if pollutants such as heavy metals can attach to these materials, their concentration will be higher in the fronts than in the surrounding sea, often by orders of magnitude. Furthermore, nutrients are pushed into frontal regions, which can form the basis for increased primary production and makes larger fronts attractive feeding grounds for fish.

One of the first observational studies of coastal plumes was reported by Blanton (1981) in the coastal current created by the many rivers that drain South Carolina and Georgia, USA. Using two ships, he mapped the salinity distribution along a section approximately perpendicular to the coast, recording the vertical structure of the currents at two stations at which the ships anchored at hourly intervals for two and a half days during a period of weak winds. The geography of the study area is shown in Figure 7.10a and an example of the vertical and across-shore density structure is shown in Figure 7.10b. Note that at both the inshore and offshore boundaries of the domain, the vertical stratification is relatively weak, and that higher values of vertical and horizontal density variations are restricted to the vicinity of stations C and D.

The classic model of a plume of fresh water emerging from a coastal river or estuary into the ocean consists of a bulge region near the mouth of the estuary. The generation of the bulge is described by the discharge Rossby number,

$$R_D = \frac{U_r}{fL_r},\qquad(7.24)$$

where U_r is the outflow velocity, f is the Coriolis parameter, and L_r is typically the channel breadth or the radius of the estuary mouth. For high R_o outflows in idealized settings, the bulge shape is circular, and a significant fraction of the freshwater discharge is diverted to a growing bulge. Low-discharge R_o outflows form small bulges, so a large part of the freshwater flows as a current trapped along the coast.

The plume is formed as a thin layer of river water over ocean water with a salinity that varies from top to bottom. The behavior is characterized by the internal Rossby radius,

$$R_{oi} = \frac{\sqrt{g\frac{\Delta\rho}{\rho_0}H}}{f}.\qquad(7.25)$$

When the plume width is small compared to the internal Rossby radius, the rotation of the Earth has little effect and the plume's motion in the coastal ocean is determined primarily by the prevailing currents. On the other hand, when the plume width is comparable to or greater than the internal Rossby radius, the Earth's rotation is important and the plume is deflected to the right of the discharge site in the Northern Hemisphere, and a coastal current of width 5–10 km is formed. The bulge region is similar in appearance to an anticyclonic gyre or eddy, with velocities approaching 50 cm/s at its circumference and slower velocities near the center. The transport around the bulge is not axially symmetric. The seaward side of the bulge has a stronger flow field than near the coast. A sketch of the plume's behavior is shown in Figure 7.11.

The plume moves about an internal radius offshore before it is affected by the offshore dynamics. Most plumes, regardless of the importance of the Earth's rotation, reattach themselves to the coast. The distance downstream depends on many factors. Figure 7.12 depicts the surface signature of the Hudson River

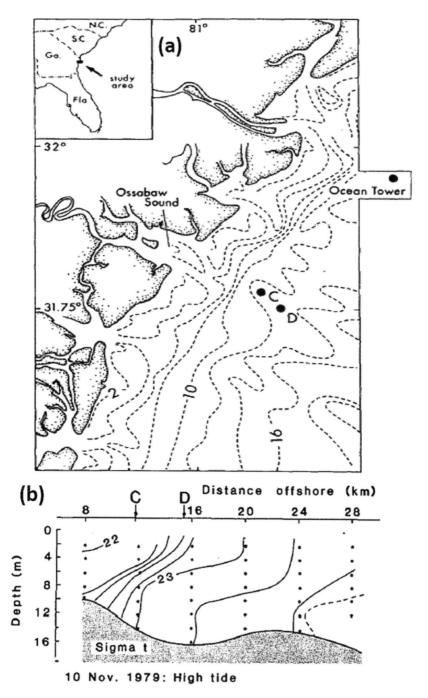

Figure 7.10 (a) The inset shows a map of the coastline and state boundaries of the southeastern United States with an arrow indicating the location of Ossabaw Sound. The more detailed map shows the coastal geometry and bathymetry of the study area with the location of anchor stations C and D. (b) A vertical cross section of the density (σ) along a line through C and D in (a). From Blanton (1981).

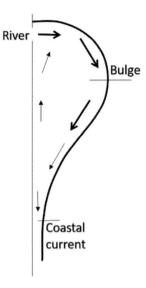

Figure 7.11 Sketch of a buoyant plume as it emerges into the coastal ocean. Here, U is the transport. Width of the coastal current is governed by the internal Rossby radius.

plume chlorophyll images from an April climatology (Figure 7.12a) and snap-shots from the 2004 (Figure 7.12b), 2005 (Figure 7.12c), and 2006 (Figure 7.12d) field seasons. The climatology resembles the classic picture of Figure 7.11 with a clearly defined bulge and coastal current. There is, however, significant inter-annual variability. In 2004, during low river flow conditions, the plume also resembles the classic picture, with a bulge and coastal current (Figure 7.12b). During the 2005 field study, which followed immediately after a nearly record-breaking river discharge, the plume forms a large recirculating bulge with little or no coastal current (Figure 7.12c).

Further offshore beyond the continental shelf are sites of major current systems, the most notable one bordering North America being the Gulf Stream. A strong steering influence of bathymetry generally keeps these currents from riding up onto the continental shelf, but there are often instabilities in these current systems that cause eddies to be shed, which impinge on the coast and may influence the cross-shelf transport. On the East Coast, instabilities of the Gulf Stream generate warm core rings and smaller eddies called shingles, which impinge on the shelf along the East Coast (Figure 7.13). They have been found to have a strong influence on the flow on the outer shelf and are likely important agents in fluid exchange between the outer shelf and the adjoining ocean. On the west coast, instabilities in the southward flowing California Current (Figure 7.14) result in a complex field of eddies adjacent to the continental shelf. These eddies

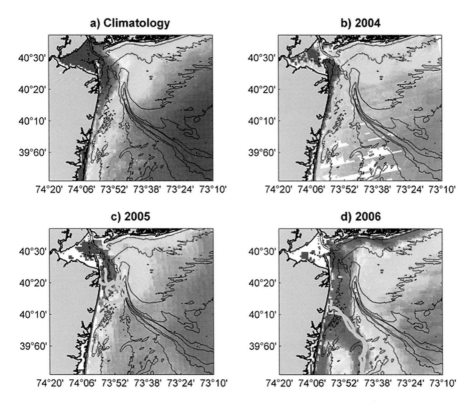

Figure 7.12 Moderate Resolution Imaging Spectroradiometer (MODIS) chlorophyll a images of the LaTTE study area for (a) April climatology (2004–2008), (b) 5 May 2004, (c) 4 April 2005, and (d) 28 April 2006. Higher chlorophyll concentrations (in red) are indicative of the presence of the Hudson River plume. Drifter deployments during the LaTTE experiments are shown in gray (Hunter et al., 2012). (A black-and-white version of this figure appears in some formats. For the color version, please refer to the plate section.)

result in strong offshore flows between the outer shelf and the ocean, which carry cold, upwelled water from the shelf into the ocean interior.

7.6   Transport of Pollutants

The typical urban coastal environment exposes the waters – including their shores and bottom sediments – to many sources of pollution. These sources include stormwater runoff, spills, septic systems, and many others. The effects of nutrient loadings from industrial and municipal wastewaters on the depletion of dissolved O_2 appears to be growing globally. The low O_2 concentrations are thought to alter the behavior, cause reduced growth, and increase mortality of

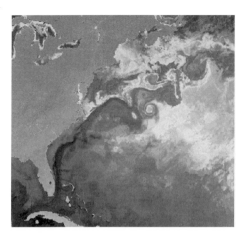

Figure 7.13 Satellite infrared image of the waters off the East Coast of the United States showing the Gulf Stream, its meandering nature and its eddies. (A black-and-white version of this figure appears in some formats. For the color version, please refer to the plate section.)

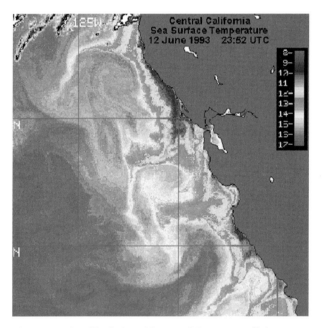

Figure 7.14 Satellite infrared image of the waters off the west coast of the United States taken June 12, 1983. The spacing of the horizontal lines is about 100 km. The upwelled coldest water is seen along the coast (Flament and Armi, 1985). (A black-and-white version of this figure appears in some formats. For the color version, please refer to the plate section.)

many marine fish and invertebrates. Many types of coastal protective features, ranging from surge barriers to natural features like wetlands and oyster beds, have been suggested as solutions for coastal flooding. Water quality and coastal flood protection are integrally linked, as flood protection efforts that reduce water circulation can in some cases harm water quality. Moreover, climate change will worsen flooding due to sea level rise and will further lower dissolved oxygen concentrations due to warming.

The introduced pollutants are transported together both vertically and horizontally by the estuarine and coastal ocean waters. Evolution of the total (dissolved and particle bound) concentration of a pollutant results from a combination of sources, sinks, and mechanical transport by a flow field. The last of these has a key role in shaping a pollutant's distribution in sea water. It has two components: transport governed by the velocity field (advection) and transport due to turbulent diffusion (see Equations 5.16 and 5.17). Turbulence in the ocean is determined by velocity gradients, surface and deep perturbations, and sea water stratification. It plays an important role in the intensity of the diffusion processes and thus a pollutant's spatial distribution. Due to advective and turbulent transport the pollutants can be readily distributed throughout an area's waters.

Observational programs have been the cornerstone of our conceptual and theoretical understanding of currents and water properties in coastal regions. Our knowledge has increased because of the introduction of moored, hydrographic, Lagrangian, and satellite-based observations. However, to permit a consistent view of the circulation, all of these types of observations are needed simultaneously. Consider, for example, that in many coastal regions the length scales of the hydrodynamical processes are characterized by a Rossby radius of 5–15 km and by topographic variations ranging from 1 to 10 km. Motions and water properties measured at stations separated by distances much greater than the length of these scales will, in general, tend to be only loosely related to one another. The sampling networks of hydrographic surveys and current meter moorings must be chosen within the Rossby radius. For coastal domains of small extent, this is possible; however, for large regions it is not always feasible. There are also a host of issues associated with obtaining observations with the proper time scales. Satellites provide excellent spatial views that are, unfortunately, only snapshots in time. Current sensing techniques are better at addressing temporal variability, since they employ sampling frequencies that are typically 30 minutes or less; however, their spatial coverage is limited. It is apparent, then, that observational programs are rarely sufficiently dense (with respect to frequency of observation) in either space or time to provide an

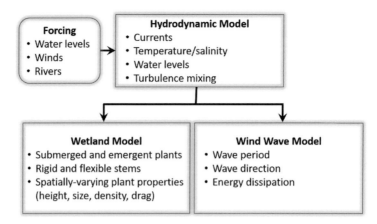

Figure 7.15 Schematic diagram of the major processes controlling pollutant fate and transport in the urban ocean.

adequate description of the water mass and velocity fields of an evolving, three-dimensional part of the coastal ocean.

In recent years, estuarine and coastal-ocean circulation models have come to be depended upon, when properly tested and verified, to synthesize information from measurements and provide a framework for investigating the basic processes of a region. In these models, differential equations, Equations 2.1, 4.23, 4.24, 4.25, 5.16, and 5.17, are solved providing water level, and three-dimensional salinity, temperature, current, and turbulence mixing distributions. A modular framework for addressing pollutant fate and transport issues in the environment is shown in Figure 7.15. The modules relate the hydrodynamics of the velocity, water levels, salinity, temperature, and turbulence, as influenced by flow–wave-vegetation and wind- and swell-driven waves.

Arguably, the most established hydrodynamic models are derivatives of the Princeton Ocean Model (POM; www.ccpo.odu.edu/POMWEB/) family. POM accounts for many sources and sinks in the various variable equations. These include time-varying river inflows, surface fluxes of momentum (rainfall), and heat fluxes through the sea–air interface. As such, models play a critical role in determining how nutrients, sediment, contaminants, and other waterborne materials are transported. A coupled wetland module is used to simulate the important flow–vegetation interactions based on a three-dimensional physics-based approach in which stem-induced drag and inertia forces are added into the mean flow momentum and turbulence transport equations. The model treats vegetation stems as cylindrical elements and takes into account the flexibility of stems by estimating the bent stem height using the finite deformation theory of a cantilever beam. The wave simulation module enables capturing vegetation drag, wave refraction, and wave-driven accumulation of nearshore and harbor water levels, a process called "wave setup". The modeling systems are available to understand the controls humans place on various discharges of municipal and industrial wastes, agricultural runoff, combined sewer overflows, waste spills of potentially toxic substances, etc. They provide a useful tool for investigating linked biogeochemical processes in the urban ocean.

8

Urban Meteorology

8.1   Introduction

Meteorological phenomena in the coastal regions are characterized by sharp changes in heat, moisture, and momentum transfers and elevation that occur between land and water. The coastal urban areas are affected by both weather and climate and exert an influence on the local scale weather and climate. The region itself spans about 100 km inland or offshore of a coastline. The climate in and around these cities and other built up areas is altered in part due to modifications humans make to the surface of the Earth during urbanization. The surface is typically rougher and often drier in cities, as naturally vegetated surfaces are replaced by buildings and paved streets. Buildings along streets form urban street "canyons" that cause the urban surface to take on a distinctly three-dimensional character. These changes affect the absorption of solar radiation, the surface temperature, evaporation rates, and storage of heat, and the turbulence and wind climates of cities and can drastically alter the conditions of the near-surface atmosphere. Human activities in cities also produce emissions of heat, water vapor, and pollutants that directly impact the temperature, humidity, visibility, and air quality in the atmosphere above cities. On slightly larger scales, urbanization can also lead to changes in precipitation above and downwind of urban areas. In fact, urbanization alters just about every element of climate and weather in the atmosphere above the city, and sometimes downwind of the city.

8.2   Urban Air Basics

The chemical composition of the atmosphere is key to understanding urban climate and weather. The Earth's atmosphere is composed mainly of nitrogen, Oxygen, and water vapor in the following percentages:

- nitrogen: 78%
- oxygen: 21%
- argon, CO_2, neon, helium, methane, others: 1%

Air is never completely dry, however, and water vapor (H_2O) can occupy as much as 4% of the air's volume. Temperature and humidity determine the density of air masses, which in turn determines whether these air masses will rise or sink. Moist air is less dense than dry air at the same temperature because the weight of water vapor (H_2O) is less than that of nitrogen (N_2) and oxygen (O_2). When water vapor increases, the amount of O_2 and N_2 decreases per unit volume.

- Molecular weight of O_2 = 16 + 16 = 32
- Molecular weight of N_2 = 14 + 14 = 28
- Molecular weight of H_2O = 1 + 1 + 16 = 18

And when heated, air expands and becomes less dense. This means that cold air is denser than warm air, and cold, dry air is much denser than warm, moist air.

8.3   Weather and Climate

A clear distinction between weather and climate needs to be made at this point. What is weather you might ask? It is what's going to happen in the atmosphere today, tomorrow, in the next few days, even a week or two. Weather deals with storms and large-scale influences. The distance scale is 1 km nearby to, let's say, 100 km, the scale of a storm itself.

Then there is the climate. Climate is what happens in the atmosphere over the longer time scale. It deals with the temperature in January as opposed to the temperature on January 10th of a particular year. Climate conversations could be, is the climate of New York City the same as the climate of Panama? No, would be the answer because Panama is near the equator and New York is much farther north. Panama's climate is always very warm and humid. In New York City, the climate depends on the time of the year. Climate is a long time-scale phenomenon while weather occurs on the short time scale. Climate is very hard to change but global warming is gradually changing the climate. Cities, we will learn, have an important impact on the weather as well as the regional climate.

8.4   Scales of Motion

Atmospheric motions occur over a broad continuum of space and time scales. The mean free path of molecules (approximately 0.1 μm) and

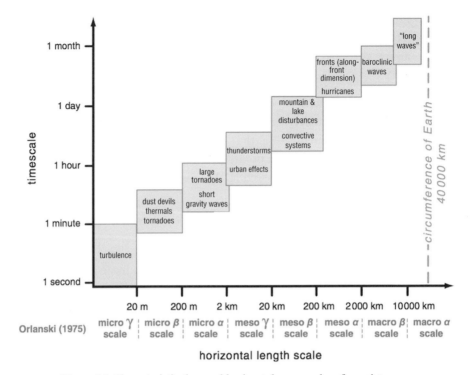

Figure 8.1 Characteristic time and horizontal space scales of a variety of atmospheric phenomena (adapted from Markowski and Richardson, 2010).

circumference of the Earth (approximately 40,000 km) place lower and upper bounds on the space scales of motions. The time scales of atmospheric motions range from under a second, in the case of small-scale turbulent motions, to weeks in the case of planetary-scale Rossby waves. Meteorological phenomena having short temporal scales tend to have small spatial scales, and vice versa. Figure 8.1 illustrates the temporal and spatial scales of some phenomena related to atmospheric motion. The largest scale, the macro-scale, includes hurricanes and nor'easters. The intermediate scale, the meso-scale, includes smaller mid-latitude cyclones, meso-cyclones, fronts, tropical storms, thermal circulations, and thunderstorms. The smallest scale, the micro-scale, is set by the size of the largest buildings and the size of a smoke stack plume interacting with tornados, cumulus clouds, turbulence, and dissipation. There is even a neighborhood scale, smaller than the micro-scale. That scale is about car exhausts, odors, dust, and pollutants. The majority of urban meteorological phenomena lie in the micro-α scale but have strong interaction with the meso-scale, typically β (Table 8.1).

Table 8.1 *Characteristic horizontal distance scale*

Scale	Macro-		Meso-			Micro-		
	α	β	α	β	γ	α	β	γ
from	Earth radius	10,000 km	2,000 km	200 km	20 km	2 km	200 m	20 m
to	10,000 km	2,000 km	200 km	20 km	2 km	200 m	20 m	↓

From Orlanski (1975).

8.5   Urban Boundary Layer

The lowest part of the Earth's atmosphere that is in direct contact with an urban setting is called the urban boundary layer (UBL). It is where all life and human activities take place in cities. The UBL typically extends about 1,000 m upward from the ground. This thickness is controlled by the wind speed, roughness of the bottom, sources of heating from the city itself, and the amount of turbulent mixing. The lowest part of the UBL is called the urban canopy layer (UCL), which extends from the ground to the average height of the building or trees in the area. From the ground, up to two to five times the height of buildings including the UCL is the roughness sublayer. Above that is the inertial sublayer and higher is the mixed layer (Figure 8.2).

During the daytime, surface heating from the sun creates large buoyant thermals that rise into the colder air (warm air is less dense than cold air), effectively carrying surface pollutants upward until they reach the top of the UBL. At that height, further vertical movement is halted by a capping inversion where there is a stable layer of cooler air below warm air blocking the vertical rise. Pollutants, heat, and humidity are trapped in the UBL, degrading the air quality and resulting in the issuance of a pollution alert, as seen in Figure 8.3. The capping inversion leads to the layer of clouds you often see from an airplane. Just below that layer the air becomes bumpy with considerable turbulence. At night, the UBL shrinks as cooling from the Earth's surface usually creates a stagnant layer near the ground about 200–400 m deep that inhibits vertical mixing into the layers above. This layer is called the nocturnal boundary layer. It is typically calm and free of turbulence. If one gets airsick, it is better to take off at night than during the day because the bottom of the atmosphere where the airplane journey begins is calmer and the airplane ride less bumpy.

8.6   The Urban Heat Island

The most obvious effect of urbanization is the trend toward higher temperatures. Figure 8.4, which is reproduced from a now classical study,

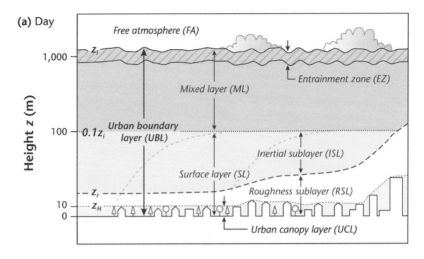

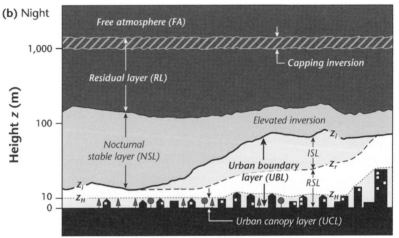

Figure 8.2 Structure of the atmosphere in an urban environment in the day (a) and at night (b). Note that the vertical scale is logarithmic (Oke et al., 2017).

illustrates a profile of air temperature from one side of New York City to the other. The rural areas are cool far away from the city. As you approach the city, the temperature increases markedly. And then as you leave the city, it gets lower again.

The warmer area shown in Figure 8.4 is called an urban heat island, an urban area that is significantly warmer than its surroundings. The temperature excess is greatest near the surface (4°C) and decreases rapidly with height. The excess on this particular day is near zero at 300 m, which is at the height of the rural UBL. In other urban areas, the temperature excess can increase up to 10°C higher than

Figure 8.3 A view of an urban boundary in Beirut, Lebanon (www.trekearth.com/gallery/Middle_East/Lebanon/West/Beyrouth/Beirut/photo99978.htm). (A black-and-white version of this figure appears in some formats. For the color version, please refer to the plate section.)

the surrounding rural areas. This temperature difference is usually larger at night than during the day and larger in winter than in summer, and it is most apparent when winds are weak. The main causes are reductions in vegetation, greater anthropogenic (from human activities as opposed to natural processes) heat emissions, along with materials used to build urban infrastructure, which store larger and larger amounts of heat. As population centers grow, they tend to change greater areas of land, which then undergo a corresponding increase in average temperature.

The effects of urban heat islands on the weather are often detrimental. Scientists have several different hypotheses that may explain how cities impact rain. One hypothesis is that the urban heat island effect, which causes warmer temperatures in cities, creates unstable air, which leads to rain. Air is unstable when it is warmer than the air around it. The warm, unstable air starts to rise. The air cools as it rises, which allows water vapor within it to condense and form clouds. If the warm, rising air is carrying enough water vapor, those clouds can grow into rain clouds. Another hypothesis is that when wind hits the skyscrapers and other tall buildings in a city, it is pushed up higher in the atmosphere. This makes unstable air. The unstable air flows upward and cools, allowing water vapor to condense, forming clouds, which can lead to rain. Air pollution in cities may also affect cloud formation and rain. Water vapor condenses on tiny particles in the air pollution, forming the droplets that make a cloud.

The heat island readily changes the flow of the larger scale meso-scale circulation as well. Weather patterns are shifted. Storms that are approaching large

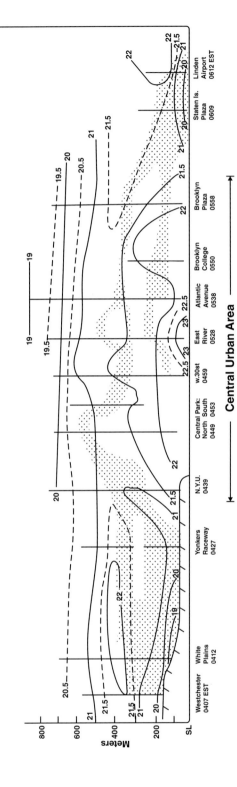

Figure 8.4 Horizontal and vertical distributions of air temperature across the New York City scape on the morning of July 16, 1964. The width of the central urban area is about 4 miles (Bornstein, 1968).

urban areas, like New York or Baltimore, from the west basically split or 'bifurcate' around the cities because of the physical structure of the buildings and because of the heat in the urban heat island. When the two halves of the storm come back together downwind of the city, the air ascends rapidly with the rising air forming rain clouds. Numerical simulations suggest a ~5–10% increase in precipitation downwind of urban regions. In cold climates, the heat island will be warmer and thus reduce the need for heating. In summer, however, the increased heat of our cities increases discomfort for everyone, and this requires a 5–10% increase in the amount of energy used for air-conditioning. And over the longer term there is increased pressure on the electrical grids leading to brownouts and blackouts.

The air quality around urban areas is also degraded. Not only are urban heat islands hotter, but they are also much more polluted. Human health becomes an issue with people experiencing respiratory difficulties, heat cramps, heat exhaustion, and unfortunately heat-related mortality.

Communities can take medium to long-term action through community planning and infrastructure to mitigate against the detrimental effects of urban heat islands. The mitigation can be accomplished in several ways; most prominent are switching dark surfaces to light reflective surfaces, by planting trees and vegetation, and by creating green roofs. Dark surfaces, such as black roofs on buildings, absorb much more heat than light surfaces, which reflect sunlight. Black surfaces can be up to 70°F (21°C) hotter than light surfaces and that excess heat is transferred to the building itself, creating an increased need for cooling. By switching to light colored roofs, buildings can use 40% less energy. An innovative initiative in Los Angeles, CA is to paint over many of the city's black roads with a reflective white coating. Adding the coating to a patch of black asphalt can keep the area up to almost 6°C cooler, a difference that could prove vital as global temperatures continue to increase.

Planting trees not only helps to shade cities from incoming solar radiation, but they also increase the evaporation of water from a plant's leaves, stem, flowers, or roots back into the atmosphere (evapotranspiration), which decreases the air temperature. Trees can reduce energy costs by 10–20%. The concrete and asphalt of our cities increase runoff, which decreases the evaporation rate and thus also increases temperature. The addition of roofs covered by vegetation help keep buildings cooler in the summer and warmer in the winter; they can store water and reduce flooding during storms and even extend the life of a roof. These green roofs can be intensive (heavy, expensive; capable of supporting large plants) or extensive (light-weight patches of shrubbery).

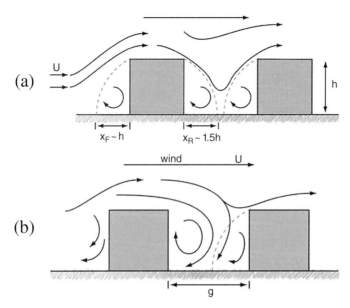

Figure 8.5 Effects of building height (*h*) and separation distance (*g*) on flow regimes (Zajic et al., 2011).

8.7   Wind Movements in the UBL

When it is windy in an urban setting, the wind can be strong in some places and almost calm in others. And it seems that the wind is coming in every direction. The winds that blow in and around the urban heat island are greatly affected by the heating and the surface roughness, i.e., the buildings and other infrastructure. The wind, which is notably retarded below the level of the buildings, shows usually a minor speed maximum just above the mean roof height. That increase is called a roof-top jet. The increase is small, on the order of 1 m/s, but during high wind events like hurricanes and typhoons, the increase for example in Tokyo was found to have a 10 m/s increase over a 30 m/s below the roof level wind.

Air flow over a regular array of identical buildings is shown in Figure 8.5. The flow between the buildings has a clockwise rotating eddy that fills the space. If the buildings are far apart, two eddies appear, and buildings whose width is about twice the building height have one eddy. For buildings that are close, the air skims over with little penetration into the space between them.

When the buildings are of unequal heights and have different separation distances, as shown in Figure 8.6, the flow patterns become quite complex depending on the windward and downwind building heights as well as the separation. Figure 8.6 illustrates how the height scale ratios $h1/h2$ and aspect

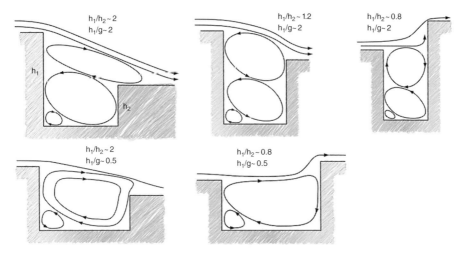

Figure 8.6 Flow patterns over buildings of unequal heights and widths. The windward height is $h1$ and the downwind building height is $h2$. The distance between buildings is g (Zajic et al., 2011).

ratio $h1/g$ affect the flow and the number of eddies. When the wind speeds are small, the thermal effects come into play; buildings tend to be warmer than the surrounding air (and depending on the intensity of heating either the approach wind or the thermal convection dominate).

Intersections play an important role in partitioning the air flow in an urban area. The flow patterns in a common four-way street intersection bounded by buildings is shown in Figure 8.7. The direction of the wind is critical in determining the wind patterns in the intersection. When the wind blows down the street, a pair of counter rotating eddies spanning a distance about the street width appear in the side streets. As the wind direction rotates counterclockwise, the response is that one of the eddies disappears and the flow becomes asymmetric and complex.

8.8   Winds Over Land vs Over the Ocean

The difference in properties of land and the ocean have significant impacts on the climate of coastal cities. It is very well known if you live along the water that the air blows from land to water and from water to land every day in the summer. This phenomenon, known as land/sea breeze, has been identified since the time of the classical Greeks (naval battle of Salamis ca. 480 BC) and was long known by fishermen as their passport to the fishing grounds at night and a safe return during morning. A land and a sea breeze is a diurnal thermally driven

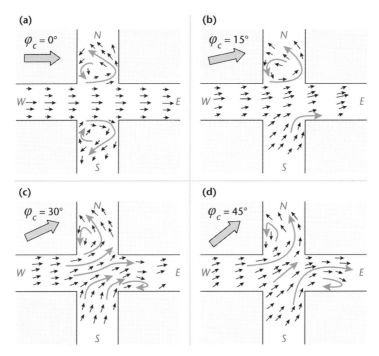

Figure 8.7 Typical flow patterns at a street intersection as a function of the wind direction, φ_c (Oke et al., 2017).

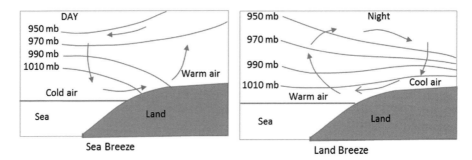

Figure 8.8 Sea breeze (left) and land breeze (right) circulation patterns set up by the temperature differences between land and the offshore ocean.

circulation in which a definite surface convergence zone exists between air streams having over-water versus overland histories.

The sea breeze shown in Figure 8.8 begins as the air over land heats up and expands more rapidly than that over water. This leads to a pressure gradient aloft that produces a horizontal pressure gradient and a slight upper level flow from the land to the sea. A convergence develops offshore leading to

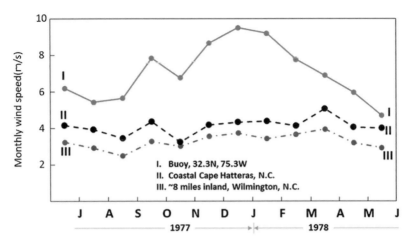

Figure 8.9 Examples of differences in monthly wind speed (m/s) at stations ranging from offshore to coastal to inland (from July 1977 through June 1978).

higher pressure and subsidence driving a flow from the sea to the land, the sea breeze. At night, the land cools more rapidly than the sea producing a pressure gradient in the opposite direction and an offshore breeze develops, the land breeze (Figure 8.8). Since the air of the sea is typically cooler than that over land, the sea breeze brings welcome relief to a warm city in summer.

The winds, whether generated by a coastal storm or an afternoon summer sea breeze, blow faster over the ocean than over the land because there is not as much friction over the water. The land has mountains, coastal barriers, trees, human-made structures, and sediments that cause a resistance to the wind flow. The oceans do not have these impediments, which impart friction, therefore the wind will blow at a greater velocity.

Figure 8.9 compares wind speeds measured at three locations: one station is inland, another on the coast, and the other is taken from a buoy location offshore. Clearly the wind speeds measured over the water are greater in magnitude than the speeds from the land stations. For many reasons related to coastal marine sciences and offshore engineering projects, a knowledge of the wind data over the water is required. Because measurements of wind over water are difficult to obtain, the tradition has been to rely on wind measurements over land, which are routinely recorded or easily obtained. Corrections to land-based wind data for offshore applications are based on simultaneous offshore and onshore wind measurements made at a host of stations ranging from Somalia, near the equator, to the Gulf of Alaska. Offshore data obtained from standard US NOAA buoys, research platforms, and merchant ships were compared with data from coastal stations. The results indicated that

$$U_{sea} = 1.62 + 1.17U_{land} \tag{8.1}$$

is sufficiently accurate (units are in m/s).

8.9   Urban Hydrosphere

The challenge of controlling and managing the water resources used by and affected by cities and urbanized communities is complex and evolving. Urbanization brings changes in land use with the construction of buildings, roads, parks, and other facilities, and increases the supply of water for consumptive use and release of wastewater. The main water problems from urbanization relate to notable changes in the urban hydrological cycle, especially the apparent increase in precipitation. Water quality becomes an issue with runoff from dirty and hot streets with heavy metals and other pollutants, and sewers that pollute local water bodies. Furthermore, urbanization means vegetation clearance, ground compaction, and introduction of impervious surface cover leading to rapid runoff of precipitation and subsequent street flooding.

Impervious surfaces and engineered drainage systems combine to drain water rapidly sometimes causing more "flash" flooding; they reduce infiltration and ground water recharge and evaporation. There is an increase in runoff since most precipitation is intercepted and cannot infiltrate or evaporate later. Stormwater becomes an issue when it rains. Where does the rain water go? It cannot infiltrate into the ground because in urban areas the ground is highly impervious, and there are few green spaces around. So, all the water goes to the stormwater drainage system, which in most older cities is the same system as the sewage disposal system (publicly owned treatment works, POTW). This single system of pipes transports both the urban runoff and sanitary sewage. During dry weather flow periods, the sanitary sewage is routed to an interceptor plant from where it is diverted to the sewage treatment plant rather than flowing to the stormwater outlet. Sometimes, during heavy rain and snow storms, combined sewers receive higher than normal flows. Treatment plants are unable to handle flows that are more than twice design capacity, and when this occurs, a mix of excess stormwater and untreated wastewater discharges directly into a city's stormwater outlet and into waterways at certain outfalls. Figure 8.10 depicts how the sewer and stormwater pipe systems work.

There is much concern about this combined sewer overflow (CSO) because of its effect on water quality and recreational uses. Sewage-polluted waters contain pathogens, heavy metals, pharmaceuticals, and chemicals that can lead to illness

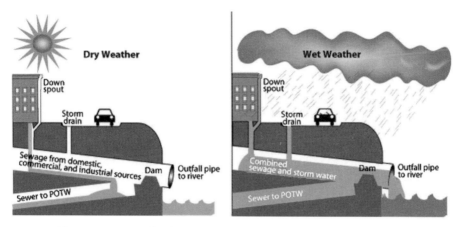

Figure 8.10 A combined sewer system in dry and wet weather (US Environmental Protection Agency, 2004).

and infection. Exposure to contaminated water from a CSO poses health risks to humans. After a CSO, bacteria such as *Escherichia coli*, viruses, and parasites can be found in the water, leading to diseases and infections. Gastroenteritis, rashes, and ear, nose and throat infections are the most common illnesses associated with contaminated water. More severe health outcomes include cholera, dysentery, infectious hepatitis, and severe gastroenteritis. Sewage and stormwater pollution are also harmful to aquatic life. In many areas, no-contact advisories are issued to tell the public to avoid contact with water in the affected area. There should be no swimming, wading, or types of water recreation or play where water could be swallowed or get in the mouth, nose or eyes. Most advisories are in effect for 5–7 days after the sewage or CSO discharge stops.

Under the Clean Water Act, cities and sewer districts can be required to bring their raw sewage discharges down to acceptable levels by reducing the frequency and magnitude of their CSOs. Right now, many cities in the United States are under mandate to reduce their CSO discharges. New York City, for example, has recently completed construction projects with upgrades in key wastewater treatment facilities, storm sewer expansions and the construction of several large CSO retention tanks to further mitigate this chronic source of pollution.

Urban floods, as we have learned, constitute one of the principal weather hazards. Loss of life and damages are often very high. The change of the surface from permeable soils and vegetation to impermeable pavements, parking lots, and roofs shortens the time it takes rain to reach the drainage system, which is often easily overwhelmed. The engineering challenges become:

1. How to control peak flows and the water levels in the drainage system to reduce flood damage at all points
2. How to predict peak flow and/or runoff volumes
3. How to design a system that will work well when the population drastically increases or when the climate changes

Much effort has been devoted to measuring the peak discharge as a function of the amount of rain and the time lag between the center of mass of rainfall excess and the peak discharge of a given watershed. Each urban drainage system has its own characteristics such as drainage area, roughness, slope, land-use, soil types, fraction of impervious surfaces, and storage characteristics. In general, the peak discharge is

$$Q = C \cdot I \cdot A, \tag{8.2}$$

where Q is the peak flow (m^3/s), C the runoff coefficient $\approx$ runoff/rainfall (a calibration parameter), I is rainfall intensity for the design storm (has to be in m/s), and A is the catchment area (m^2). The runoff coefficient is defined as the ratio of runoff to rainfall and is a function of watershed characteristics including land use, soil type, and slope of the watershed. The value of runoff coefficient ranges between 0.0 and 1.0. A value of 0.0 means that all of the rainfall is lost to infiltration, interception, and evaporation, and none of the rainfall is converted to runoff. The value of 1.0 implies that all of the rainfall is converted to runoff and is discharged from the watershed. To provide a better sense of Equation 8.2, runoff coefficient values for various land uses, soil types, and slope conditions are provided in Table 8.2.

Urbanization has seriously aggravated precipitation-driven flooding but there are considerable measures being taken now to control urban runoff. The measures are mainly focused on best management practices (BMPs) such as low-impact development (LIDs) or green infrastructure, pervious pavements, detention and infiltration basins, bioretention systems, and constructed wetlands. A very promising concept is the idea of stormwater basins. They are an excavated area designed to collect the stormwater to prevent flooding or erosion and are classified as dry and wet basins (Figure 8.11). The dry basin, the detention pond, is used mainly to prevent peak flow erosion through creating a temporary space for the water to reside for a short period of time, usually around 24–48 hours. It is normally covered in dry grass, except during storm events. Therefore, it can be multifunctional such as used for parks or playgrounds during its dry period. The wet basin, also called a retention pond, is used mainly to prevent flooding and downstream erosion by retaining the water for a longer time. It has a permanent pool of water, which is only exceeded when it rains. It promotes pollutant removal through

Table 8.2 *Runoff coefficient values*

Description of area	Runoff coefficient
Business	
Downtown	0.70–0.95
Neighborhood	0.50–0.70
Residential	
Single-family	0.30–0.50
Multi-units, detached	0.40–0.60
Multi-units, attached	0.60–0.75
Residential (suburban)	0.25–0.40
Apartment	0.50–0.70
Industrial	
Light	0.50–0.80
Heavy	0.60–0.90
Parks, cemeteries	0.10–0.25
Playgrounds	0.20–0.35
Railroad yard	0.20–0.35
Unimproved	0.10–0.30
Character of surface	
Pavement	
Asphaltic and concrete	0.70–0.95
Brick	0.70–0.85
Roofs	0.75–0.95
Lawns, sandy soil	
Flat, 2%	0.05–0.10
Average, 2–7%	0.10–0.15
Steep, 7%	0.15–0.20
Lawns, heavy soil	
Flat, 2%	0.13–0.17
Average, 2–7%	0.18–0.22
Steep, 7%	0.25–0.35

Nicklow et al. (2006)

sedimentation or absorption by the aquatic vegetation. This pond offers an opportunity for landscape designers to create an aesthetic element for the city. It also can be used for wildlife habitat or as part of a public open space.

8.10   The Future

The accelerating growth of coastal urban populations, themselves, on a very small fraction of the Earth's surface area, have become a driving force of

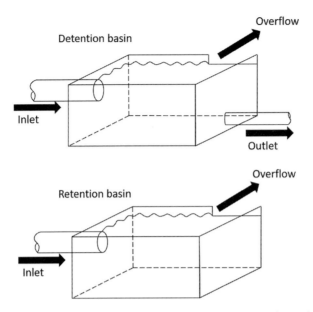

Figure 8.11 Schematic of the differences between detention and retention basins (adapted from DeGroot, 1982).

human development. The urban areas are important sites for greenhouse gas emissions because of the high energy demands by urban residents and activities. These emissions extend the influence of cities on climate to much larger scales, affecting the weather even on regional scales. Locally altered urban climates that have existed for many years may provide us with some insight into how to respond to large-scale climate change, and this makes the study of urban climates increasingly important.

The advances in urban observational techniques can provide better observations required to address the future of the urban coasts. Some relevant points to this topic are:

- Continued and expanded urban model intercomparisons over urban areas
- Development and application of best practices to strengthen the dialog between meteorologists and end user communities
- Increased access to observational data in different categories from diverse sources
- Creation of capabilities for integrated urban meteorology–decision support systems.

9

Coastal Processes and Shoreline Modification

9.1 Introduction

When we think of the coast, most of us think of long stretches of sandy beach, perhaps dotted with vacation homes, resorts, marinas, and even a casino or two. Very few of us would think in terms of a constantly evolving landscape, a "soft" boundary between land and ocean that very likely looks quite different – and may even be in a different location – than was the case 100 or even 50 years ago! But whether we are speaking about an undeveloped coastline or the highly urbanized coastline that is the focus of this book, we must begin the discussion with a recognition that the coastline is in fact one of the most dynamic regions on Earth. It is a geological feature of our planet that is shaped by influences from the ocean side and the land side, from the various sources of sand that include riverine sediments and relic coral, to the various forces that re-work the sand once it gets to the coast, including ocean waves and currents, and the wind. It is also frequently and profoundly shaped by humans, via the introduction of structures and processes that interfere with the natural movement of sand to, from, and along the coastline.

As an example of this complex interplay among the forces of nature and humans, Figure 9.1 illustrates the evolution of the coastline at Atlantic City, New Jersey from 1841 to 1948. Note the location of the boardwalk along the northern portion of the coastline in 1886, and the Absecon Lighthouse, noted as "Absecon L. H." just landward of the 1886 boardwalk. The lighthouse now stands more than two city blocks landward of the coastline, which became actively stabilized by the government in the early 1900s in order to protect the boardwalk, hotels, and amusements that lined the shore at that time. The stabilization projects included the Oriental Ave jetty at the Absecon Inlet,

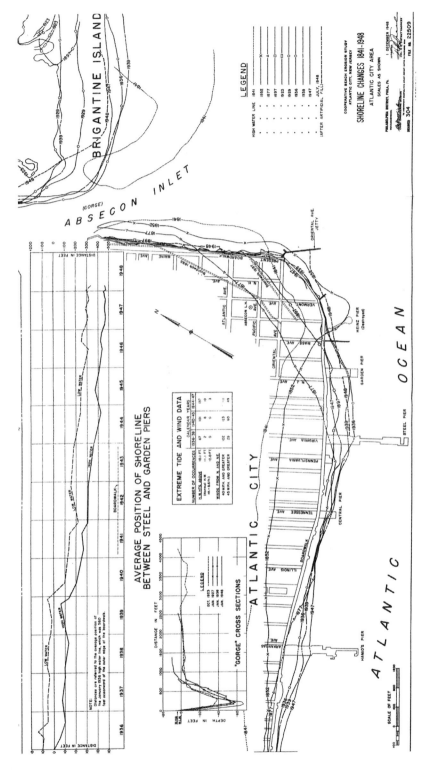

Figure 9.1 Location of the coastline at Atlantic City, New Jersey from 1841 to 1948 (US Army Corps of Engineers, 1948).

several sand-trapping beach-perpendicular stone structures, known as groins, along the beach, and the occasional addition of sand to the beach, particularly after major storm events.

In this chapter, we will discuss the processes that govern the creation and evolution of coastal shorelines, including both long-term sedimentation and short-term extreme events. Human influences such as the introduction of coastal structures, the elimination of natural protective features (e.g., wetlands), and the interference with natural sand supply and sand movement on both the water side (e.g., navigation channel dredging) and the land side (e.g., river dams) are discussed here in much more detail than was discussed in prior chapters. The impacts associated with climate change and sea level rise are also discussed.

9.2   Coastal Wave Dynamics

Figure 9.2 illustrates the basic characteristics and nomenclature associated with an idealized surface gravity wave. We refer to surface ocean waves as "gravity" waves because the restoring force (the force seeking to create a flat water surface) is gravity, as opposed to, e.g., the force due to surface tension that serves as the restoring force for very small capillary waves. We refer to the

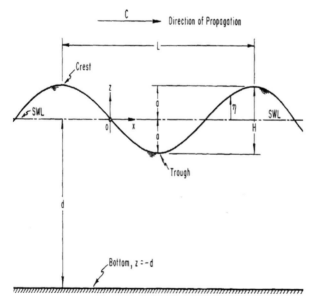

Figure 9.2 Idealized surface gravity wave, propagating from left to right (adapted from US Army Corps of Engineers, 1989).

waveform in Figure 9.2 as "idealized" because it is defined by a sinusoidal, propagating wave:

$$\eta = a \cos\left(\frac{2\pi x}{L} - \frac{2\pi t}{T}\right), \tag{9.1}$$

where:

η = the position of the water surface in space (x) and time (t)

a = wave amplitude

L = wave length, often written as wavelength, and defined as the distance from crest to crest (or trough to trough)

T = wave period, defined as the time it takes for the wave to travel crest to crest (or trough to trough)

In Figure 9.2 we also define a few other important features of the waveform and environment:

C = wave speed, often referred to as the phase speed, and defined as L/T

H = wave height, defined as $2a$

d = the water depth

SWL = the "still water line," which defines the water surface if no waves were present.

The water motion under a wave is elliptical and varies greatly with distance from the water surface, as shown in Figure 9.3, with higher, circular velocities near the surface and lowest, purely horizontal velocities near the bottom. We must remember that the wave in Figure 9.3 is propagating from left to right, and that the water velocity under a wave is a dynamic property. The forward motion occurs under the crest of the wave, and the reverse motion occurs under the trough of the wave.

As waves approach the shoreline, they begin to "feel" the bottom, that is to say the presence of the seafloor begins to affect the wave dynamics. Importantly, the wave begins to slow down. As a result, two distinctly different but important phenomena occur: shoaling and refraction. Shoaling describes the transformation of the wave into a form that differs dramatically from the idealized waveform in Figure 9.2. The wave crest becomes taller and narrower while the wave trough becomes lower in its vertical extent but more elongated. Eventually, at the location where the water depth, d, equals approximately $0.78\,H$, the wave breaks, as depicted in Figure 9.4. This is the location where the highest amount of energy is released into the water.

Refraction of the incoming waves occurs because waves never approach the shoreline at exactly 90 degrees to the coast. One can imagine therefore that an approaching wave first feels the bottom at the shallowest, or landward-most,

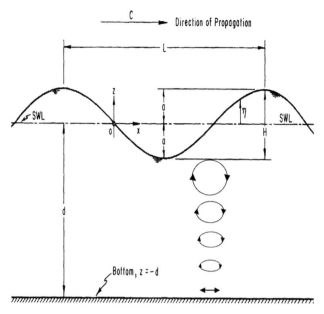

Figure 9.3 Pattern of water motion under an idealized surface gravity wave, propagating from left to right (US Army Corps of Engineers, 1989).

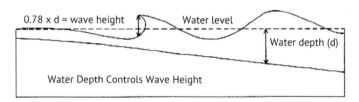

Figure 9.4 Wave shoaling and breaking.

point. This region of the wave slows down while the deepwater portion of the wave continues at its original forward speed. As a result, the wave slowly "bends", or refracts, so that the wave direction becomes more and more perpendicular to the coast. This is illustrated in Figure 9.5.

9.3   Properties of Sediments

The sand (sediments) that make up our beaches can vary widely in their origin and in their physical characteristics. This is important because the properties of the sediment will govern the eventual configuration of the beach and its response to coastal storms and human activity.

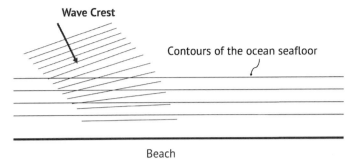

Figure 9.5 Wave refraction.

Coastal sediments are derived primarily from three sources: land-based rocks, volcanoes, and biologically derived material. Land-derived, or "terrigenous", sediments are the result of the chemical or physical transformation ("weathering") of larger, land-based rocks. This material is introduced to the ocean via rivers, glaciers, and wind, and is often composed of quartz, although other material (e.g., Calcite and Dolomite) can be present. Volcanic sediments are delivered to the ocean by the wind, as ash and dust from volcano eruptions. Although these sediments are found in larger quantities nearest to volcanic islands, they can be found all around the Earth, as they can be transported long distances by the winds, especially during and after large-scale eruptions that can cause plumes to rise many thousands of feet into the atmosphere. Large-scale ocean currents also transport this material to distant locations on the ocean floor. Biologically derived, or "biogenic," sediments are associated with biological activity. They are primarily the result of the chemical or physical transformation of shell fragments, skeletal remains of plankton, coral, etc. These sediments are composed of calcium carbonate, because these animals are calcite-secreting organisms. Silica-secreting organisms such as sponges and diatoms also contribute to biogenic sediments. These sediments are composed of silicon dioxide. Of these two dominant contributions to biogenic sediments, the "carbonate" sediments are roughly three times by weight more abundant than the "siliceous" sediments. It is important to note that the sediments found along any particular coastline are often not composed of any one of the three source-types described here but are rather a combination of material contributed by several different sources over many, many years.

Sediment grain size is characterized by the diameter of the grain. Table 9.1 provides the size range of the various "classes" of sediment, ranging from very fine clay to very large boulders.

In addition to the size of the sediment grains, another property that is important to the deposition and movement of ocean sediments is the density of the material. Table 9.2 lists the density of the most common sedimentary materials.

Table 9.1 *Size range of various classes of sediment*

Size range (mm)	Wentworth classification
4,096 to 256	Boulder
To 2	Cobble, pebble, granule (often referred to as gravel)
To 1	Very coarse sand
To 0.5	Coarse sand
To 0.25	Medium sand
To 0.125	Fine sand
To 0.0625	Very fine sand
To 0.0039	Silt
To 0.0005	Clay

Adapted from US Army Corps of Engineers (1998).

Table 9.2 *Density of various sedimentary materials*

Mineral	Density (kg/m^3)
Quartz	2,648
Feldspar	2,560–2,650
Kaolinite	2,594
Calcite	2,716
Aragonite	2,931
Dolomite	2,866

From US Army Corps of Engineers (1998).

9.4   Physics of Sediment Motion

Once sedimentary material is deposited on the seafloor or on the bed of a river or estuary, its movement is governed primarily by the size of the grains of sediment and the dynamics of the water body. Sediment grains are thrown into the water ("suspended") by turbulence generated by, e.g., waves and are thereby made available to be transported by the current; they are also rolled along the bottom by the near-bottom water motion. The complex physics of sediment transport in ocean and estuary waters is beyond the scope of our discussion, but in Figure 9.6 we provide an illustration of the various processes responsible for the motion of sediment within the water column. Note that sediment moves both along the bottom (as "bed load transport") and within the water column (as "suspended load transport"). The type of transport that dominates in any particular instant is a complex function of the sediment grain properties (primarily the size and weight of the grains) and the water velocity and turbulence in the water column. Energetic environments, such as might be expected in the shallow

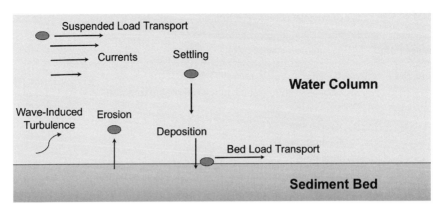

Figure 9.6 Sediment transport processes.

waters of a coastline with wave action, would likely give rise to the erosion of sand grains from the seafloor and subsequent movement of the sediment as suspended load transport.

Clearly, a key question to be answered when examining the sediment transport dynamics of a coastal location is whether or not the sediments at that particular location are mobilized by the observed waves and/or currents. Figure 9.7 provides a useful tool for the assessment of conditions for no transport (deposition), or transport via either suspended or bed load transport. This is a modified version – developed by Madsen and Grant (1976) – of the well-known Shields Diagram, which was developed by Shields (1936), who employed laboratory measurements to develop an understanding of the relationship between the initiation of sediment movement and the water motion and sediment properties. The "Shields Parameter" is defined as:

$$\psi = \frac{\tau_0}{(s-1)\rho g d},$$ (9.2)

where:

τ_0 is the shear stress at the bottom, which is proportional to U_b^2, where U_b is the near-bottom water velocity

ρ_s is the sediment density, ρ is the density of the water, and s is the ratio of ρ_s/ρ

g is the acceleration due to gravity and d is the sediment grain diameter

In Figure 9.7, v is the kinematic viscosity of the water. Values above the curve indicate sediment movement. In essence, the figure indicates that for a particular grain of sediment, increased bottom shear stress, or water velocity, will likely produce sediment grain motion. This motion is more likely to be

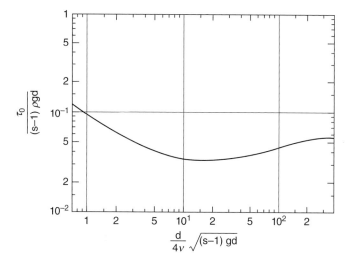

Figure 9.7 Modified Shields Diagram (Madsen and Grant, 1976). Values above the line indicate that the sediment is moving.

suspended motion for smaller grain sizes and bottom motion (think of this as rolling) for larger grain sizes.

One important note needs to be made regarding the erosion, movement, and deposition of coastal sediments. This is regarding the behavior of "cohesive sediments". These sediments lie in the size range corresponding to silt and clay, from 0.0005 to 0.0625 mm. Sediments having grain sizes in this range are affected by an additional force – the attractive force, or cohesive force, associated with the electrochemical forces between charged particles. This attractive force is greatest in salt water and gives rise to a process called "flocculation" wherein the sediment grains stick together to form larger agglomerations of grains, known as "flocs." These flocs are much more readily deposited on the seafloor. However, once they settle on the sediment bed they are bound together as cohesive layers that become quite resistant to erosion. These types of erosion-resistant sediment beds can also be created by bottom-dwelling organisms that secrete mucus, causing the fine-grained sediment to glue together. Many coastal environments, in particular those regions with tributaries and estuaries that drain large watersheds, contain a mix of cohesive and cohesionless (sand and larger) sediments.

9.5   Coastal Sediment Transport

Sediment that is deposited along coastal ocean regions, whether derived from rocks, volcanoes, or biological activity, and whether deposited via rivers, ocean currents, glaciers, or wind, are mixed and transported by an assortment of

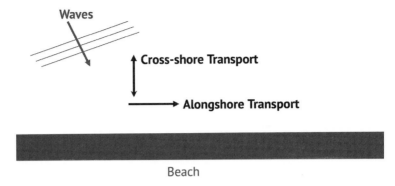

Figure 9.8 Definition of the two modes of coastal sediment transport.

water motions and currents caused by the combined action of waves and currents. In its simplest form, the transport can be broken down into two components: cross-shore transport and alongshore transport, as illustrated in Figure 9.8.

Traditionally, coastal scientists and engineers have chosen to treat each mode of sediment transport on its own. As one can imagine however, the erosion, suspension, and transport of sediment is a very complex phenomenon, with the possibility of sediment traveling in a "zig-zag" pattern along the coast, moving onshore and offshore while also moving alongshore. But let's here treat each mode separately.

9.5.1   Cross-Shore Sediment Transport

Cross-shore transport is primarily driven by wave action. As discussed earlier in this chapter, as waves approach the shoreline, they begin to shoal. As the wave crest becomes more narrow and high, the onshore motion associated with the crest becomes stronger, but short-lived. Meanwhile, the offshore motion associated with the lower but longer trough becomes weaker, but longer-lived. As a result, sediment is moved forward in short bursts under the crest and offshore for a longer duration under the trough. Of course, as has already been discussed, the movement of the bottom sediment is determined by both the water velocity and the sediment characteristics, primarily the grain size and weight. One interesting outcome of this complex physics is that a natural "sorting" can occur on the beach. Only the larger velocities under the crest can mobilize larger grain sediments. Although the finer-grained sediments are also mobilized under the crest, these are also mobilized under the trough, but for a longer duration. As a result, the grain size of the bottom sediments will typically increase as one moves toward the beach.

Perhaps the most dramatic outcome of cross-shore sediment transport is the "bar-berm" response of the beach profile. As illustrated in Figure 9.9, the beach

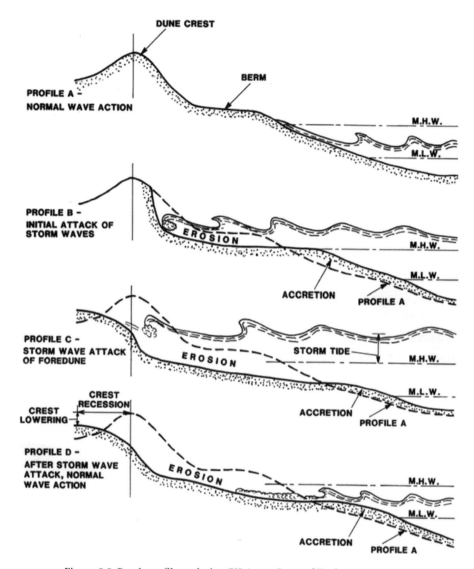

Figure 9.9 Beach profile evolution (US Army Corps of Engineers, 1984).

profile (cross-section) on a natural shoreline often includes a landward sandy area of dunes or a coastal bluff, extending landward until the limit of extreme wave run-up is reached. In populated coastal regions, this natural sandy area is often replaced with man-made structures such as seawalls, intended to protect valuable property or critical infrastructure. Moving toward the ocean, the beach berm is the dry, sandy beach that is often used for recreational purposes. As inferred in Figure 9.9, the beach profile is in constant flux. The most dramatic evolution occurs during periods of high wave action such as during a storm. The breaking waves will erode and suspend the bottom sediments as mentioned earlier, making the sediment available for motion. This motion is dominated by offshore-directed motion, caused by the longer-lived offshore motion under the wave troughs but also more importantly, the strong return current produced as the water that is "piled up" along the beach returns toward the ocean. This return current follows a path of least resistance and flows along the bottom, where the wave-generated currents are the weakest, as mentioned earlier. This offshore-directed sediment motion continues until the conditions are favorable for sediment deposition (a consequence of the combination of sediment characteristics and water motion, as depicted in Figure 9.7). As time goes on, an equilibrium condition is reached where the beach profile has fully adjusted to the storm conditions, typically resulting in an eroded beach berm and dune, and an offshore accretion (deposition) area, often referred to as a sand "bar." In a very real sense, the presence of this sand bar helps to reduce and even prevent further erosion of the beach, as the large storm waves begin to break further offshore rather than expending their energy on the beach berm and dune. This response of a natural beach to severe wave conditions represents a natural form of storm protection and shoreline preservation.

During subsequent, post-storm conditions having normal wave action, the beach would be expected to return to something like the original profile. That is, unless the offshore movement of sediment is so dramatic that the normal wave action is not able to return the sand to the beach berm, as can occur during very severe storms.

We have here defined this evolution of the beach profile in the context of "normal" and storm wave conditions. We should note that in most regions of the world, the ocean wave conditions vary dramatically by season, with the winter season characterized by storm waves and the summer season characterized by mild wave action. As a result, many if not most shorelines exhibit a seasonal cycle of beach profile shape, with the "summer profile" resembling Profile A (sometimes referred to as a "berm profile") in Figure 9.9, and the "winter profile" resembling Profile D (sometimes referred to as a "bar profile").

9.5.2   Along-Shore Sediment Transport

In addition to moving onshore and offshore in response to wave action, sediment is moved along the coast, or "alongshore" by wave action and/or the combined influence of waves and currents. As mentioned earlier in the context of wave refraction, waves approach the coastline at an oblique angle. This results in the generation of an alongshore current, something that any swimmer has experienced at the beach if they try and remain at the same location while treading water. As might be expected, given that the current is generated by the incoming waves, the magnitude of the current is highest roughly in the region where the waves are highest, that is, near the area where the waves are breaking.

The influence of this alongshore current, often referred to as the "longshore" current, is remarkable. Longshore, or "littoral", sediment transport can move many tons of sand per day along a given section of beach. Since the direction of the transport varies entirely with the wave direction, this component of sediment transport can fluctuate on a daily basis. Over a long period of time, however, most coastlines have a dominant direction of incident wave action, resulting in a dominant direction of longshore sediment transport. The results of this transport are often easy to recognize, sometimes to a staggering degree. Figure 9.10 is an aerial view from the mid-1990s of a section of the northern New Jersey coastline, just south of New York Harbor. In this area of the coast, the dominant wave direction is from the south. As a result, the longshore sediment transport is on average directed toward the north, and sediment accumulates on the southern side of coastal structures, as is evident at the bottom of the photo. Even more dramatic is the major geological feature of this region of the coast, the sand "spit" known as Sandy Hook, visible in the upper portion of the photo. This sand landmass has built up over the years and formed a miles-long landform that contains a popular park and historic buildings dating to the early 1800s. Sandy Hook also forms the southern "gateway" into New York Harbor, with the harbor entrance channel lying just to the north of the landform.

9.5.3   Beach Erosion

As we have already discussed, the beach profile, and with it the position of the shoreline, evolves across time scales that can vary from hours (e.g., during storms) to months (e.g., seasons), and even longer. The loss of beach width, normally described as beach erosion, is therefore a natural response of the coastal system to the various natural forces at work. Although the word "erosion" has very negative connotations, it is only in the context of human development that this connotation exists. If it were not for homes and infrastructure placed in harm's way, there would be no problem!

Experience has shown, in fact, that even in developed coastal regions, if the coastline is allowed to evolve without interference from humans, negative impacts associated with beach erosion are largely absent, except for the possible erosion caused by a change in ocean dynamics, e.g., via sea level rise. Of course, this is increasingly a rare situation, as it requires that a populated coastline be left to evolve naturally, with dune systems (normally vegetated) fronted by a

Figure 9.10 Photo of Northern New Jersey beach with Sandy Hook in the background, ca. 1995 (photo by M. Bruno). (A black-and-white version of this figure appears in some formats. For the color version, please refer to the plate section.)

berm, both of which are continuously replenished via wind and ocean-borne sediment sources. In order to delve deeper into this topic, it is useful to consider beach erosion across two very different time scales, one – short-term erosion – associated with time scales of hours and the other – long-term erosion – associated with time scales of decades.

9.5.3.1   Short-Term Erosion

Short-term erosion can be defined as the rapid loss of beach sediments to the ocean, and landward migration of the shoreline in response to severe coastal storms. Most commonly, this erosion occurs over time periods ranging from a few hours (e.g., during tropical cyclones) to multiple days (e.g., during extratropical storms). Very often, and via the processes already discussed, the erosion experienced during a sort-lived storm will be reversed during subsequent quiescent periods when mild wave conditions return the offshore sand to the beach berm. However, severe storms, such as occurred during Hurricane Katrina in 2005 and Hurricane Sandy in 2012, can produce permanent loss of sediment and beach width. This is in part because the sediment is removed to deep water from which subsequent wave action cannot return it to shore (recall that wave-induced motion decreases with distance from the surface). If the eroded beaches lie in areas that have been influenced greatly by human development, particularly with respect to the availability of new sediment sources, the storm-induced erosion can be catastrophic. Examples of such human influences include the interruption of longshore sediment transport by the construction of shore-perpendicular structures such as inlet jetties, the trapping of river-borne coastal sediments by dams and bulkheads, and the replacement of dune systems with seawalls and other infrastructure. Often, the short-term erosion and shoreline change produced in these areas during extreme storm events can exceed – in less than 48 hours – the long-term erosion over many decades.

The combined influences of a severe storm and a human-altered beach profile is dramatically illustrated in Figure 9.11, which provides two snapshots of the erosion of the beach berm in front of a seawall along the south shoreline of Long Island, New York. This erosion occurred as a result of a series of three storms in the Fall of 1997. Prior to the first storm, much of the seawall and the concrete stairs down to the beach were covered with sand. It is remarkable to see the vertical extent to which the beach berm has been eroded. The wall collapsed soon after the photo on the right was taken.

Figure 9.12 illustrates the possible consequence of the combined effects of a major storm and human development along a high-risk coastal region, in this case the devastating impact of Hurricane Sandy in October 2012 along the

Figure 9.11 Accelerated beach erosion in front of a seawall in Long Island, NY, 1997 (photos by M. Bruno). (A black-and-white version of this figure appears in some formats. For the color version, please refer to the plate section.)

Figure 9.12 Aftermath of Hurricane Sandy at Mantoloking, New Jersey, October, 2012 (photo courtesy of US Army Corps of Engineers). (A black-and-white version of this figure appears in some formats. For the color version, please refer to the plate section.)

developed coastline of northern New Jersey. The area is characterized by a "barrier beach", meaning a relatively thin stretch of beach that separates the ocean on one side from a bay on the other. Situations such as this are very complex, because the storm surge associated with the tropical storm influences the ocean and the bay (via inlets) in different ways. Importantly, the timing of the peak of the surge on the ocean side and the bay side can be several hours apart.

As a result, many hours of flooding can occur. In the case of Hurricane Sandy, the record level of flooding and the severely eroded beach berm and dune allowed the water from the bay to flow out and into the ocean. During subsequent ocean high tides, the surge and wave-induced water motion pushed massive amounts of water toward the bay, carving out a new inlet and removing numerous beachfront homes.

Although it would be reasonable to conclude that the construction of homes on this narrow strip of sandy beach prevents the beach from responding to a major storm in a natural way, the human influence extends beyond that. In fact, the primary cause of erosion in this region, as in many other developed coastal regions, lies in a deficit of sediment supply to the beach. As mentioned earlier, the interruption of longshore sediment transport can occur as a result of many human activities, including the construction of shore-perpendicular structures such as inlet jetties, the trapping of river-borne coastal sediments by dams and bulkheads, and the replacement of dune systems with seawalls and other infrastructure. Along the coastal region illustrated in Figure 9.12, all of these influences are present. When combined with a storm of the magnitude of Hurricane Sandy, it is no wonder that the natural system sought an entirely new balance in the dynamics between the ocean, bay, and the beach.

9.5.3.2   Long-Term Erosion

Long-term erosion and shoreline change are a complex result of a multitude of factors, including of course not only the cumulative effect of storms, but also sea level rise, and human impacts such as the interruption of sediment supply and transport, as has already been discussed in the context of short-term erosion.

A dramatic example of the long-term erosion produced by the reduction or elimination of sediment supply to the coast is the construction of the Aswan High Dam on the Nile River in Egypt, which was completed in 1970. The dam is enormous, at just over 12,500 feet long and over 360 feet high. Historical records indicate that the maximum flow rate along the Nile River downstream from the Aswan High Dam was reduced by anywhere from three to as much as seven times, and the suspended sediment concentration (the most important indicator of sediment supply) dropped from 3,800 ppm (parts per million) to 129 ppm during flood events (Rasian and Salama, 2015). It has been estimated that the total sediment volume delivered to the coast from the Nile River has been reduced from 60 to 180 million tons per year to less than 15 million tons per year (Torab and Azab, 2006). According to the authors' analysis of shoreline position data, the delta region of the Nile River in the vicinity of the western ("Rosetta") branch of the river experienced dramatic erosion in the years following completion of the Aswan High Dam. The long-term average erosion rate increased from approximately 13.7 to 20 m per year in the period from 1900 to 1964 to approximately 95.3 to 124.8 m per year in the period from 1964 to 2006!

As discussed earlier, severe storms often carry sediment far enough offshore that subsequent wave action cannot return the sediment to the beach. This sediment is for all practical purposes lost to the coastal transport system. Over time, a succession of even moderate storm events can produce a permanent loss of beach width and hence, shoreline recession.

9.5.3.3   Impact of Sea Level Rise

Sea level rise is also causing gradual beach erosion and shoreline recession along a majority of coastlines around the world. As water levels rise, waves approach closer to the shore (remember the relationship of breaking wave height to depth). This places more of the beach berm and perhaps even the dune at risk of erosion. The beach profile adjusts to this new water level by moving landward, causing a recession in the shoreline position. The worry of course is that along developed coastlines, there is little or no room for the beach berm and dune to migrate landward!

The exact relationship between sea level rise and shoreline recession is very complex and highly variable from location to location. It is not simply a linear relationship between the water level and the beach slope, which might predict, for example, a 20-feet recession with a 1-foot sea level rise for a beach with a slope of 20 feet horizontal to 1 foot vertical. The reality is that the 1-foot sea level rise produces new dynamics higher on the beach. Along coastlines with complex topography, e.g., barrier beaches, the increased sea level affects the dynamics on both sides of the beach.

Figure 9.13 illustrates the global sea level from a combination of tide gauges along the coasts and altimeters on satellites, as measured from the global average sea level in 1990. Tide gauge measurements since the early 1900s indicate that global sea level rose at an average rate of 1.7 mm per year, or approximately two-thirds of an inch per decade. Satellite altimeter data from 1993 to 2003 indicate that the rate of global sea level rise increased to 3.1 mm per year, or about one and a quarter inches per decade. The altimeter record is still too short to be certain whether we are observing an acceleration of global sea level rise, but the information should be of concern to coastal communities worldwide.

Before moving on from this topic, it is important to recognize that while global average sea level is rising, the actual position of the sea level *relative to a point on land* is highly variable. Figure 9.14 illustrates the relative sea level rise at

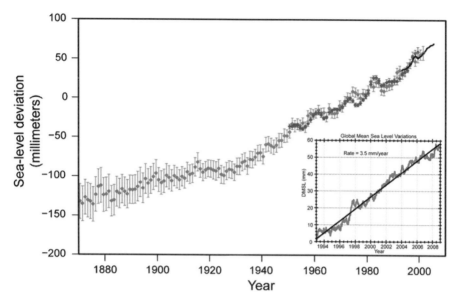

Figure 9.13 Annual averages of global sea level. The red line indicates the trend in sea-level since 1870. The dark line in the primary figure illustrates the measurements based on tide gauge data, with the last portion based also on satellite observations. The inset shows the sea level rise measured since 1993, during the period in which satellite measurements were available (IPCC, 2007). (A black-and-white version of this figure appears in some formats. For the color version, please refer to the plate section.)

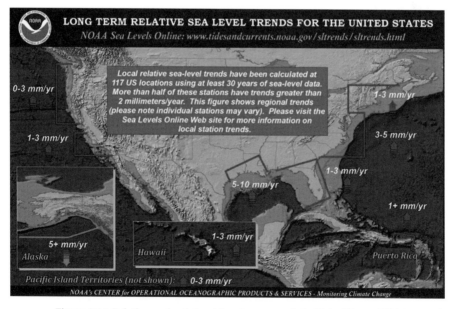

Figure 9.14 Relative sea level rise at locations around the United States (NOAA, 2015a).

various locations around the United States. Note that the rates vary from as low as near zero in the Pacific Northwest to 10 mm per year (4 inches per decade) along some areas of the Gulf Coast. Much of this variability is due to the combined effects of sea level rise and land subsidence, with human activities, such as the removal of underground water and oil and coastal land development, playing a sometimes major role.

10

Marine Pollution

10.1 Introduction

In the previous chapters an account has been given of the chemical and physical aspects of the elements that together constitute the urban ocean environment – namely, (1) the seawater itself and (2) the water movement dynamics. The chemical constituents and the physical properties of the seawater, together with their distribution, concentrations, and cyclic changes, and the movement of this water, are the governing factors in the history and fate of a fascinating array of living things.

The ecology of urban coastal regions demonstrates the special challenges associated with concentrated human activity. In the waters around New York City, for example, sharks were considered the greatest danger of being in the waters until the latter half of the nineteenth century. The city was considered the "capital of American angling," attracting sharks at the top of the food chain to the waters. Reliable accounts spanning more than a century, from 1760 to 1881, show that the Manhattan waterfront and the harbor were often infested with sharks (Figure 10.1). The sharks are all gone now. What happened to New York's sharks? The environmental history of New York Harbor portrays just awful conditions – oil-coated waters that caught fire and portions of its bottom were covered in raw human sewage sludge many, many feet thick. Not surprisingly, many fishes, oysters, and herons and other birds of prey were only a memory. New York Harbor has also experienced profound physical alteration since the Colonial era. This includes dredged channels, filling of wetlands, creation of artificial islands, construction of piers and sea walls, and the blasting of reefs hazardous to navigation (see Waldman, 2013). The New York City experience is just one of many such stories. This chapter will develop an understanding of the delicately balanced maintenance of life in urban coastal areas.

Figure 10.1 Sharks in New York Harbor circa 1880 (Zeisel, 1990).

10.2　Sources of Pollutants

By far the greatest volume of waste entering the urban ocean is composed of organic material, which degrades over time due to an oxidative process that ultimately breaks down the organic compounds to stable inorganic compounds such as CO_2, and ammonium (NH_4). These wastes include:

- Oxygen-demanding waste
  - Organic debris & waste + aerobic bacteria
  - Source: Sewage, feedlots, paper-mills, food processing
- Organic chemicals
  - Oil, gasoline, plastics, pesticides, solvents, detergents
  - Sources: industrial effluent, household cleansers, runoff from farms and yards

Agricultural fertilizers have a similar effect to the organic wastes. Nitrates and phosphates enter in waters as runoff and are dispersed into the urban waters by currents and turbulence mixing. The fertilizers enhance phytoplankton production, sometimes to the extent that decaying and dead plants lead to anoxic conditions. Excess nitrogen causes explosive growth of toxic microscopic algae, poisoning fish and marine mammals. Wastes in this category include:

- Plant nutrients
  - Nitrates, phosphates,
  - Source: sewage, manure, agricultural and landscaping runoff

There are many industrial discharges into the urban waters that rapidly lose their damaging properties after they enter the water. Most effects are limited to the area immediately around their point of discharge, although the extent of that area depends on the currents and turbulence mixing of the area.

Power plants, for example, take in water from rivers, heat it up as part of the plant's operations, and then eject that water back into the natural resource, which changes the oxygen levels and can have disastrous effects on local ecosystems and communities. Construction sites add sediments that are washed into waterways, choking fish and plants, clouding waters, and blocking sunlight. The wakes from ships and ferries now so common to the urban waters have the potential to cause environmental damage in the vicinity of wetlands and low-energy coastlines where wake-waves can cause extensive shoreline erosion, resuspend and transport bottom sediments, and harm aquatic wildlife.

The "pollutants" in this category include:

- Eroded sediment
  - Soil, silt
- Heat/thermal pollution
  - Source: power plants, industrial
- Boat traffic
  - Waves, wakes
  - Source: increased shipping

There are some materials that do not degrade over time and are not subject to oxidative processes that would turn them into stable inorganic compounds. These materials are called conservative wastes. They often react badly with plants and animals leading to harmful effects including cancer of several types. They are regarded as a very serious threat to humans and the environment. The principal categories of such wastes include:

- Inorganic chemicals
  - Acids, metals, salts
  - Sources: surface runoff, industrial effluent, household cleansers
- Radioactive materials
  - Iodine, radon, uranium, cesium, thorium
  - Source: coal & nuclear power plants, mining, weapons production, natural

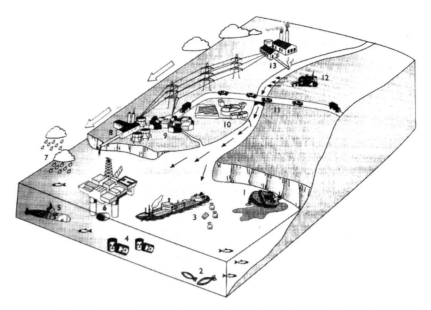

Figure 10.2 Pollutant pathways in the marine environment from a large variety of sources, including: (1) oil spills; (2) lost or dumped munitions; (3) garbage and waste from ships; (4) dumped nuclear and industrial waste; (5) lost or dumped vessels, their cargoes, and power plants; (6) oil drill cuttings; (7) washout of atmospheric pollutants including heavy metals and hydrocarbons; (8) industrial wastes and street drainage; (9) urban wastes and street drainage; (10) sewage effluent; (11) traffic exhausts (via the atmosphere); (12) agricultural fertilizers and pesticides; (13) waste heat from cooling water (Clark, 2001).

Marine debris and litter are among the largest threats facing our oceans. Plastics are used in an enormous and expanding range of products due to their relatively low cost, ease of manufacture, and versatility. Most are petroleum-based plastic, a product designed to last forever. They pose an ever-increasing problem to aquatic environments, as they do not biodegrade. Plastics break down into smaller and smaller pieces, but do not get absorbed into our natural systems and therefore never disappear. "Pollutants" in this category include:

- Plastics and litter
  - Land based
  - Offshore sources

The discharging of industrial, nuclear, sewage, and many other types of waste into the urban ocean was legal in the United States until the early 1970s when it became regulated under the Clean Water Act; however, discharges still occur illegally everywhere. Pathways that bring pollutants into the waters are from a large variety of sources. Figure 10.2 illustrates the most important ones.

In the urban ocean, the major types of pollution are excess nutrients, marine debris, oil spills, and toxic contaminants. They will be the focus of the next several sections.

10.3   Nutrients

Nutrients come from a variety of different sources. They can occur naturally as a result of weathering of rocks and soil in the watershed but many stem from human activities. Scientists are most interested in the nutrients that are related to people living in the coastal zone because human-related inputs are much greater than natural inputs. Because there are increasingly more people living in coastal areas, there are more nutrients entering our coastal waters from wastewater treatment facilities, runoff from land in urban areas during rains, and from farming.

Wastewater (or sewage) treatment plants are point sources of nutrients by virtue of the effluent that they discharge directly to the urban waters. Unless the effluent has received tertiary treatment, or treatment to remove nutrients, it can be a significant contributor. Septic systems may contribute large amounts of nutrients, particularly if located close to the water. Standard septic systems do not remove nitrates; however, special systems like sand filters that remove nutrients are now becoming more common.

Atmospheric nitrogen comprises about 78% of the air that humans breathe. The burning of fossil fuels forms oxidized nitrogen compounds, which then reach the Earth when it rains or snows. In some parts of the United States, the Northeast and the Upper Midwest, the so-called acid rain associated with these processes conveys large nitrogen loads to lakes and streams. The US Geological Survey estimates that more than 3.5 million metric tons (nearly 7 billion pounds) of atmospheric nitrogen are deposited in the United States each year.

Commercial fertilizers useful for agriculture are a major source of both phosphorus and nitrogen. According to the US Geological Survey, about 12 million metric tons (26 billion pounds) of nitrogen and 2 million metric tons (4 billion pounds) of phosphorus are applied annually in commercial fertilizer in the United States. Depending on the composition of the soil in an area, irrigation amounts and application methods, and the amount of rainfall, nutrients not needed by crops either run off the land into lakes and streams, build up in the soil, or seep down into groundwater. Groundwater can seep into a stream and be a source of nutrients.

Nitrogen and phosphorus play a major role in stimulating primary production by plankton in the oceans. These elements are known as limiting because plants cannot grow without them. However, there are a number of other nutrients that also play a role, including silicon, iron, and zinc. Nutrients in the ocean are cycled by a process known as biological pumping, whereby plankton extract the nutrients out of the surface water and combine them in their organic matrix. Then when the plants die, sink, and decay, the nutrients are returned to their dissolved state at deeper levels of the water column. The abundance of nutrients determines how fertile the oceans are. A measure of this fertility is the primary production, which is the rate of fixation of carbon per unit of water per unit time. Primary production is often mapped by satellites using the distribution of chlorophyll, which is a pigment produced by plants that absorbs energy during photosynthesis.

Under the right conditions, including abundant nutrients, algae and aquatic plants will continue to grow and multiply well beyond the amount needed to support the food web. The excess growth then dies, and microorganisms break it down, consuming dissolved oxygen from the water in the process. Dissolved oxygen, which aquatic organisms need just as humans need oxygen from the air, can be completely used up by the breakdown process. When this happens, aquatic organisms die from lack of oxygen. Extensive fish kills can result. The process of enrichment of urban waters with nutrients and the associated biological and physical changes is called eutrophication. It is a natural process, but human activity has dramatically increased its rate in many waterbodies. Areas with low dissolved oxygen are often referred to as a "dead zone" because most marine life either dies, or, if they are mobile such as fish, leave the area. Dead zones occur in many areas of the Unites States, particularly along the East Coast, the Gulf of Mexico, and the Great Lakes, but there is no part of the world that is immune. The second largest dead zone in the world is in the Unites States, in the northern Gulf of Mexico; the first largest is in the Black Sea, but that is a natural occurrence.

Given the advances in hydrodynamic and water quality modeling and advances in computing power in recent years, sophisticated modeling systems exist to address the eutrophication issue in a rigorous scientific fashion. Three-dimensional hydrodynamic models (see Section 7.6) are coupled with three-dimensional water column eutrophication submodels with biogeochemical variables; phytoplankton; multiple forms of carbon, nitrogen, and phosphorus; dissolved oxygen; and fecal coliform. The water column is coupled to sediment diagenesis models that compute sediment-water fluxes of dissolved oxygen, methane, ammonium, nitrate, and phosphate, based on computed inputs of particulate organic matter (Figure 10.3).

The modeling systems when properly calibrated and validated provide a powerful tool in water research and management to develop an understanding of the inter-relationships between nutrients, primary production, water column

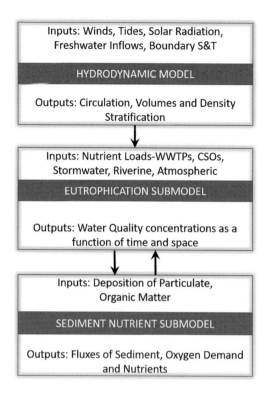

Figure 10.3 Components of a modern eutrophication model.

stratification, and low levels of bottom water dissolved oxygen (hypoxia). By understanding the processes that influence and control concentrations of bottom water dissolved oxygen, water quality managers may be able to determine required levels of nutrient reductions required to reduce the temporal and spatial extent of hypoxia, thereby improving habitat for bottom-dwelling organisms.

10.4   Marine Debris

Most marine debris (80%) comes from trash and debris in urban runoff, i.e., land-based sources. Key components of land-based sources include litter, trash and debris from construction, ports and marinas, commercial and industrial facilities, and trash blown out of garbage containers, trucks, and landfills. Ocean-based sources, such as overboard discharges from ships and discarded fishing gear, account for the other 20%. Food containers and packaging are the largest component of the municipal solid waste stream. These items, together with plastic bags, also represent the largest component of marine debris (that is, barring items less than 5 mm such as pre-production plastic pellets, fragments,

Figure 10.4 Marine debris at the beach (NOAA, 2018c).

and polystyrene pieces). Packaging and single use disposable products are not only ubiquitous in marine debris, they represent an unsustainable use of precious resources (oil, trees, energy sources, water).

Litter and marine debris in waterways pose a major threat to marine life, with the major impacts being ingestion or entanglement. However, litter and debris are also hazardous to humans, creating toxic waterways, carrying invasive species, depleting fisheries, requiring expensive clean-up operations, and ruining the aesthetics of the environment (Figure 10.4).

Unlike naturally based paper, plastic is extremely slow to degrade. Table 10.1 indicates the approximate time it takes for garbage to decompose in the environment. Plastic bottles take 450 years to decompose. They eventually break down into smaller and smaller pieces forming microplastics. Larger pieces of plastic floating at the surface are readily mistaken for food by seabirds and turtles, while plastic bags and fishing lines can wrap around marine life and kill them. Throughout the world, around one million seabirds and 100,000 marine mammals are killed every year by plastics, either entangled and strangled or choked and starved.

Although marine plastic pollution is becoming more widely known by consumers, businesses, and governments, much of the attention is currently focused

Table 10.1 *Approximate time it takes for garbage to decompose in the environment*

Glass bottle	1 million years
Monofilament fishing line	600 years
Plastic beverage bottles	450 years
Disposable diapers	450 years
Aluminum can	80–200 years
Foamed plastic buoy	80 years
Foamed plastic cups	50 years
Rubber-boot sole	50–80 years
Tin cans	50 years
Leather	50 years
Nylon fabric	30–40 years
Plastic bag	10–20 years
Cigarette butt	1–5 years
Wool sock	1–5 years
Plywood	1–3 years
Waxed milk carton	3 months
Apple core	2 months
Newspaper	6 weeks
Orange or Banana Peel	2–5 weeks
Paper towel	2–4 weeks

New Hampshire DES (2018)

on macroplastics – larger pieces of over 5 mm in size. Macroplastic pollution poses severe, well-publicized problems for marine fauna. Over 200 marine species are known to suffer from entanglement or ingestion, which can lead to choking and physical blockages, malnutrition, strandings, and even death.

But perhaps a more important aspect of marine plastic is that of microplastics (particles measuring less than 5 mm). One of the biggest problems with micro-plastic particles is that they are difficult to contain once they enter the marine environment because of their small size and ability to float. Once they reach the sea, the surface of these microplastic particles often become colonized by micro-organisms, which alter the properties of the plastic, causing it to sink through the water column and become embedded in the seabed, shoreline, and plant matter. Clean-up operations thus become labor intensive, time consuming, and costly.

Additionally, seabed filter feeders such as lugworms and mussels have been proven to consume microplastic particles, which can cause circulatory blockages. Not only is this harmful for the animals themselves, but there are also human health concerns about eating mussels contaminated with microplastics. The effects of microplastic consumption are not just restricted to filter feeders either.

Nurdles (small plastic pellets about the size of a pea) resemble floating fish eggs and are regularly mistaken as a food source by a multitude of marine fauna. Seabirds are particularly affected. For example, it is estimated that 95% of northern fulmars contain microplastics in their stomachs. In all cases, once consumed, these particles can lead to physical blockages, malnutrition, choking, and even starvation. Several studies have shown some seabirds even regurgitate microplastics to their young during feeding.

Many groups and individuals are active in preventing or educating about marine debris. The latest, launched 4 October 2017, is the ONE OCEAN Forum (see www.oneoceanforum.org/), forming an international dialog on ocean sustainability, creating a network of "intelligence" and identifying best practices to raise awareness about the issues affecting the marine ecosystem. It is organized by Yacht Club Costa Smeralda and Princess Zahra Aga Khan, in collaboration with UNESCO's Intergovernmental Oceanographic Commission and SDA Bocconi Sustainability LAB.

The Marine Pollution Convention (known as MARPOL), Annex V, is an international treaty that regulates the disposal of garbage aboard ships. Here garbage includes "all kinds of food, domestic and operational waste, excluding fresh fish, generated during the normal operation of the vessel and liable to be disposed of continuously or periodically." Annex V also prohibits the discharge of plastics anywhere in the sea (see www.imo.org/en/OurWork/Environment/PollutionPrevention/Garbage/Pages/Default.aspx). Ships, according to the treaty, must keep a garbage logbook to track all disposal and incineration aboard the ship. National governments must also provide facilities at ports and terminals to collect garbage from ships. It is illegal to throw plastic into waters within 200 miles of the US coastline (which constitutes the country's Exclusive Economic Zone). It also outlawed the dumping of garbage within 3 miles of shore.

Three important Acts enacted in the United States are:

> The Shore Protection Act created regulations for waste transport vessels like trash barges. The act aimed to prevent accidental spills of dangerous waste into the water.
>
> The Coral Reef Conservation Act. This act benefits coral reefs by authorizing NOAA to "provide assistance to States in removing abandoned fishing gear, marine debris, and abandoned vessels from coral reefs to conserve living marine resources."
>
> Marine Debris Research, Prevention, and Reduction Act. This law funded NOAA's Marine Debris Program to "identify, assess, reduce, and prevent marine debris and its effects on the marine environment."

10.5   Oil Spills

Crude oil naturally occurs in the environment originating from fossil sources, created over millions of years from deposits of microscopically small marine organisms, mainly diatoms. It is a complex mixture of hydrocarbons (carbon-based compounds with hydrogen atoms attached). Petroleum hydrocarbons are the primary constituents in oil, gasoline, diesel, and a variety of solvents. They do not generally mix with water in the urban ocean but float on its surface. Some hydrocarbons are known to be potent carcinogens. Those are called polycyclic aromatic hydrocarbons (PAHs) and are in a class of chemicals that occur naturally in coal, crude oil, and gasoline. They also are produced when coal, oil, gas, wood, garbage, and tobacco burns.

Oil pollution is one of the most conspicuous forms of damage to the marine environment. Oil enters the seas not only because of spectacular oil rig disasters like the Deepwater Horizon incident in the Gulf of Mexico but also – and primarily – from diffuse sources, such as leaks during oil extraction, illegal tank-cleaning operations at sea, or discharges into the rivers that are then carried into the sea. It is difficult to precisely estimate global oil inputs into the marine environment because the sources are so diverse. Around 5% comes from natural sources, and approximately 35% comes from tanker traffic and other shipping operations, including illegal discharges and tank cleaning. Oil inputs also include volatile oil constituents, which are emitted into the atmosphere during various types of burning processes and then enter the water. This atmospheric share, together with inputs from municipal and industrial effluents and from oil rigs, accounts for 45%. A further 5% comes from undefined sources.

When liquid oil is spilled on the sea surface, the oil immediately forms a thin film and into large slicks that float and spread on the water's surface. The rate of spreading, and therefore the thickness of the thin film, depends on the sea temperature, nature of the oil itself, and the currents prevailing in the region of the spill. A proportion of the oil evaporates or sinks, but other oil constituents are broken down by bacteria or destroyed by solar radiation. This breakdown is influenced by various physical, chemical, and biological processes (Figure 10.5). Depending on a variety of different environmental conditions, such as temperature, nutrient content in the water, wave action etc., the breakdown of the petroleum hydrocarbons may take shorter or longer periods of time. During the first few hours or even during the first few weeks, the oil is modified by the following chemical and physical processes:

- evaporation of volatile constituents
- spreading of the spilled oil in large oil slicks drifting on the surface waters

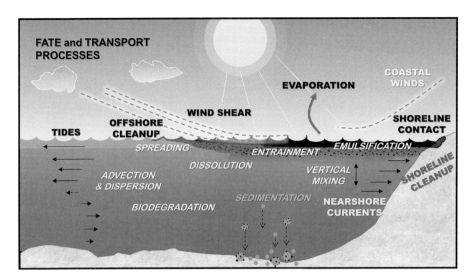

Figure 10.5 Processes illustrating how oil is modified and broken down in the sea (courtesy of Michael Fichera).

- formation of dispersions (small oil droplets in the water column) and emulsions (larger droplets of oil-in-water or water-in-oil)
- photooxidation (molecular changes to the oil constituents caused by natural sunlight) and solution

Finally, the oil solidifies into clumps (tarballs), which are more resistant to bacterial breakdown.

Processes such as sedimentation and breakdown by bacteria, on the other hand, may continue for months or even years. The speed of breakdown depends primarily on the molecular structure of the oil constituents and by the rate at which the various hydrocarbons are broken down by bacterial activity. The effects of oil spills on humans and animals may be direct and indirect, depending on the type of contact with the oil spill.

Exposure to oil spills occurs where people and animals come in contact with oil spill components:

- By breathing contaminated air – since petroleum products have many volatile compounds that are emitted as gases from spilled oil, the air becomes contaminated with those volatile oil products or vapors producing specific odors. Once in the air, contamination may travel over long distances.
- By direct contact with the skin – people and animals may come in direct contact with oil and/or oil products while walking and swimming in a contaminated area.

- By eating contaminated food – some oil compounds bioaccumulate in living organisms and may become more concentrated along the food chain. Humans may become exposed to concentrations of contaminants in the food that could be orders of magnitude higher than in the contaminated environment. This is especially problematic since residents could be exposed even if they live far away from an oil spill if they consume food coming from a spill affected area.

Although more oil is being transported in larger and larger supertankers, technical, political, and legal expertise in managing the problem has been gained mostly through conventions initiated by the International Maritime Organization and by the United States passing the Oil Spill Act of 1990. Concern by industry and citizens over the cost and damage from spilled oil have greatly increased the rate of spill reporting. Overall, oil spillage is down. Oil spill trends from 1980 to 2002 indicate that the average annual volume of oil spilled decreased 44%, and the annual number of spills greater than 500 gallons decreased by nearly 50%. Still, an average of 8 million gallons of oil is spilled annually in the United States alone.

Models have been developed to understand and predict the possible fate and behavior of oil that is introduced to the urban ocean. Information on the hydrodynamics and the waves (Figure 7.15), wind speed and direction, and oil type each with its own characteristics are the main data input to the models (Figure 10.6). The oil droplets are mathematically treated like individual particles and transported by the currents and turbulent diffusivities using Lagrangian (refer back to Section 4.3) particle tracking methods. The basic processes affecting the fate of the oil itself are taken into account and parameterized in the oil spill processes modules (Figure 10.5). In general, a caveat with the oil spill models has to do with their accuracy. Despite a number of oil trajectory and fates models being available around the world, few developers have carried out the analysis to quantitatively address the skill of the predictions.

The oil spill models are widely used for contingency planning where they can be particularly helpful for decision-making. By modeling a series of the most likely oil spill scenarios, decisions concerning suitable response measures and strategic locations for stockpiling equipment and materials can be made. The locations shown to be the most vulnerable can be identified, the logistics of responding to these locations studied, and response equipment placed accordingly. Modeling is particularly useful for ecological risk assessment. Modeling allows quantification of potential impacts and probabilities of those impacts. The relative impacts of various spills can be used to focus response efforts. Maximum liabilities for accidental spills may be estimated. The results of various

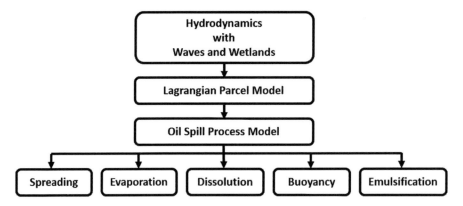

Figure 10.6 Components of modern oil spill fate and transport models.

management strategies may be investigated. A model system may be used to educate the public about potential impacts of various spill scenarios.

10.6   Metals

Heavy metals are natural components of the Earth's crust. They cannot be degraded or destroyed. They can become contaminants when industrial activity concentrates them at higher than normal levels. Since they are elements, they are not subject to bacterial attack like oil and cannot break down into anything else. Metals released from mining and industrial processes are among the major contaminants of concern in urban ocean environments, where they can accumulate in sediments and coastal organisms. The major contaminants are mercury, cadmium, copper, zinc, chromium, and silver.

To a small extent metals enter our bodies via food, drinking water, and air. As trace elements, some heavy metals (e.g., copper, selenium, zinc) are essential to maintain the metabolism of the human body. However, at higher concentrations they can lead to poisoning. Heavy metal poisoning could result, for instance, from drinking contaminated water (e.g., lead pipes), high ambient air concentrations near emission sources, or intake via the food chain. Heavy metals are dangerous because they tend to bioaccumulate. Bioaccumulation means an increase in the concentration of a chemical in a biological organism over time, compared to the chemical's concentration in the environment. Compounds accumulate in living things any time they are taken up and stored faster than they are broken down (metabolized) or excreted.

In the urban oceans, the most problematic heavy metal is Mercury. Mercury is by far the most toxic of the metals. It is a toxic substance that has no known function in human biochemistry or physiology and does not occur naturally in living organisms. Inorganic mercury poisoning is associated with tremors, gingivitis, and/or minor psychological changes, together with spontaneous abortion and congenital malformation.

Mercury is a global pollutant with complex and unusual chemical and physical properties. The major natural source of mercury is the degassing of the Earth's crust, emissions from volcanoes, and evaporation from natural bodies of water. Worldwide mining of the metal leads to indirect discharges into the atmosphere. The usage of mercury is widespread in industrial processes and in various products (e.g., batteries, lamps, and thermometers). It is also widely used in dentistry as an amalgam for fillings and by the pharmaceutical industry. Concern over mercury in the environment arises from the extremely toxic forms in which mercury can occur.

Mercury is mostly present in the atmosphere in a relatively unreactive form as a gaseous element. The long atmospheric lifetime (of the order of 1 year) of its gaseous form means the emission, transport, and deposition of mercury is a global issue.

Natural biological processes can cause methylated forms of mercury to form, which bioaccumulate over a million-fold and concentrate in living organisms, especially fish. These forms of mercury, monomethyl mercury and dimethyl mercury, are highly toxic, causing neurotoxicological disorders. The main pathway for mercury to humans is through the food chain and not by inhalation.

In recent years, there have been decreases in metal pollution, but the reductions have been slow to come by. The best way to reduce metal pollution is by reducing inputs in the first place and remediating areas that are already contaminated. In 2013, the United Nations Environment Programme (UNEP) finalized the Minamata Convention on Mercury. Major highlights of the Minamata Convention include a ban on new mercury mines, the phase out of existing ones, the phase out and phase down of mercury use in a number of products and processes, control measures on emissions to air and on releases to land and water, and the regulation of the informal sector of artisanal and small-scale gold mining. The Convention also addresses interim storage of mercury and its disposal once it becomes waste and sites contaminated by mercury as well as health issues.

The European Union has moved aggressively to combat mercury exposure. A ban on mercury exports began in 2011, and the Union has issued rules on storing mercury and restrictions on some products containing mercury, like thermometers. It is currently considering additional rules on mercury in dental fillings and batteries. Beginning in January 2012, the United States banned the export of elemental mercury, whose uses include gold mining. The new policy results from the Mercury Export Ban Act of 2008.

10.7   Pathogens

Waterborne pathogen contamination in water resources and related diseases are a major water quality concern throughout the world. The coastal urban waters typically contain pathogens that derive from human excrement as well as from other animals. However, it does not follow that swimming at a beach, or swallowing the water, will automatically cause a person to become ill. One must first encounter the pathogens, have the pathogens gain entry to the body, and have a dose high enough to overcome the body's natural defense mechanisms. The higher the concentration of pathogens, the greater the probability that health problems will develop. Even in an advanced nation like the United States, where sanitary procedures and waste disposal practices are far superior to those used in emerging economies, waterborne disease outbreaks are not uncommon. Presently in the United States, about 2000 illnesses and 10 to 12 outbreaks per year are attributed to waterborne microorganisms. This is 10% of the number of incidents that occurred prior to the development of public sewers and water treatment systems.

There are practical and scientific limitations on the extent to which water can be treated or monitored even in advanced nations. Clearly examining the nature of the waterborne pathogen problem and methodologies available for treatment and monitoring is essential for creating vibrant coastal urban communities. Most human pathogens can be divided into several classes including:

> *Bacteria* – these tiny (most range in size from 0.2 to 10 microns, 0.0000079 to 0.00039 inches) single-celled organisms are present in the bodies of all living creatures, including humans. Bacteria play a vital role in processes such as decomposition and digestion. Bacteria can be found in large numbers in raw sewage, effluents, and in natural waters. Some well-known diseases caused by pathogenic bacteria include cholera, dysentery, shigellosis, and typhoid fever.
>
> *Viruses* – too small to be recognized by ordinary light microscopes, viruses are generally recognized by the symptoms that they produce in the host. All viruses are parasites and must grow on living tissue. Many viruses are associated with feces and are expected to be found in domestic wastes. Because many viruses can survive for extended periods of time in natural waters and can occasionally withstand the treatment process, they pose a public health concern. Viruses of concern that are transported in water include hepatitis A, Norwalk-type virus, rotavirus, and adenovirus.

Protozoans – these single-celled organisms can grow up to 5 mm long and are found almost entirely in aquatic environments. Pathogenic protozoans compose approximately one-third of the entire class and can cause serious health problems such as gastrointestinal disease, dysentery, and ulceration of the liver and intestines.

It is important to keep in mind that seawater typically contains about 1 billion bacteria and 1 trillion viruses that are not pathogenic to humans.

Pathogens found in human excrement, whether urine or fecal material, come from persons who are presently infected by the disease organism. There are some persons that are infected by a pathogen who do not know it because they have no symptoms. Pathogens may enter waters through point and non-point sources, while others may occur naturally in the environment. Some point sources are wastewater treatment facilities and combined sewer overflows. Non-point sources include land and road runoff, human sewage from recreational boats, and septic systems. The major sources of bacterial contamination are due to the point sources. Treatment facilities have greatly reduced the number of pathogens that are released into the environment through disinfectant processes. However, treatment is not always 100% effective and breakdowns in facilities sometimes occur. Wildlife, domestic animals, and birds may also contribute pathogens to the environment.

The detection of pathogens is key to the prevention and identification of problems related to health and safety. Since it is impossible to test waters for every possible disease-causing organism, measures for fecal coliforms and other sewage indicator bacteria are put into place. The presence of bacterial indicators suggests that the water may be contaminated with untreated sewage and pathogenic bacteria or viruses may potentially be present. A criterion based on the indicator bacteria concentration is used to determine if waters are safe for human use. Bacteria from the fecal coliform group are used as indicators because they are not usually present in unpolluted waters and are easily detected by simple laboratory procedures. These routine tests utilized for pathogen identification, however, do not directly characterize virulence factors. Thus, these tests do not provide the needed information about the potential pathogenicity or virulence of the identified organisms.

Despite the real need for obtaining analytical results in the shortest time possible, traditional and standard bacterial detection methods may take up to 7 or 8 days to yield an answer. This is clearly insufficient, and many researchers have recently geared their efforts toward the development of rapid methods. The advent of new technologies, namely biosensors, has brought in new and promising approaches. However, much research and development work is still needed before biosensors become a real and trustworthy alternative.

The main goal is to reduce the number of pathogens that are being discharged into the coastal waters. To reduce the concentrations of microorganisms that are released, most cities are working to abate the combined sewer overflows. In addition, municipalities are repairing and better maintaining their aging sewage collection systems to ensure proper treatment. Pump-out facilities are also being constructed to help the problem of sewage from marine toilets on recreational boats from being discharged directly into the coastal waters. Domestic animals are not allowed on beaches from spring to fall when recreational areas are in high use. In addition, feeding seagulls and other seabirds is not permitted in these areas.

10.8   Epilogue

For centuries, it was thought that the urban waters were so vast and absorbing that there was nothing that people and industries could do that could possibly have an impact on them. Wrong. The typical urban environment exposes their waters – including their shores and bottom sediments – to many sources of pollution. These sources include industry, municipal wastewater, stormwater runoff, spills, vessels, septic systems, and many others. In addition, the dense urban populations adjacent to these waters add pollution from households, ranging from cleaning products to paints, and motor oil to fertilizers. Taken together, these pollutants cause many problems that are unique to urban waters: dirty shorelines and beaches; contamination of shellfish-growing areas; water unsafe for contact recreation; and toxic substances that enter the water supply and the food chain. Cleanups are expensive to business and taxpayers, and polluted areas often cannot be used for development, industry, or recreation without costly remediation. People who live near the coasts are a part of the problem and the solution to ocean pollution. Through education, they can be informed of the types of pollution and actions that they can do to prevent further pollution of the ocean.

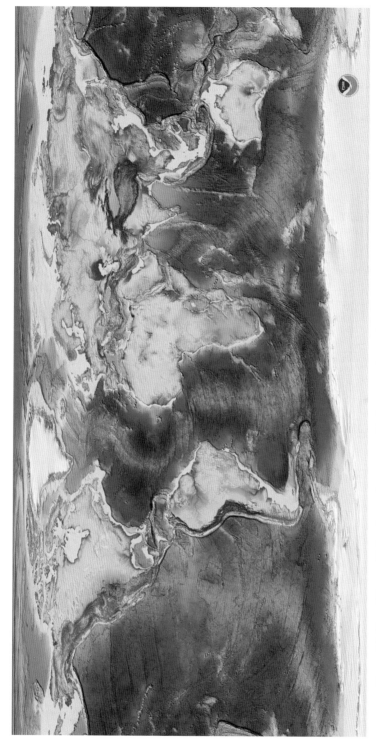

Figure 1.1 Map of the world showing major topographic features. The continental shelves are indicated with the color cyan. Note the very wide continental shelves along the coastlines of Argentina, northeast North America, Southeast Asia, northern Australia, and the Arctic (Amante and Eakins, 2009).

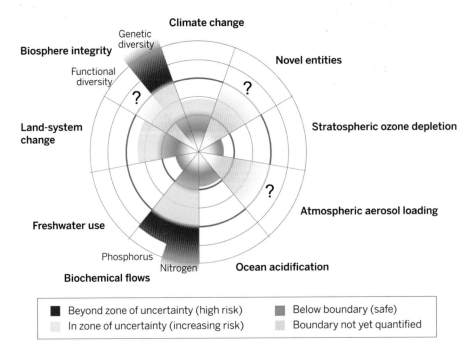

Figure 1.3 Planetary Boundaries. Reproduced from Steffen et al. (2015), with permission.

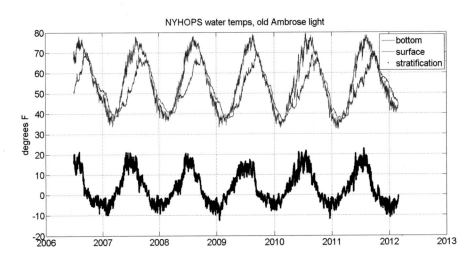

Figure 2.2 Surface and bottom temperatures and the temperature stratification.

Figure 3.1 View of the coastline in the vicinity of Honolulu, on the island of Oʻahu, Hawaiʻi (NOAA, 2018a).

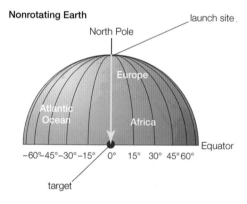

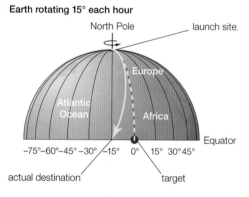

Figure 4.2 Path of a rocket launched toward the Equator from the North Pole showing that the projectile would land to the right of its true path (Reprinted with permission from the Encyclopedia Britannica, © 2008 by Encyclopedia Britannica, Inc.)

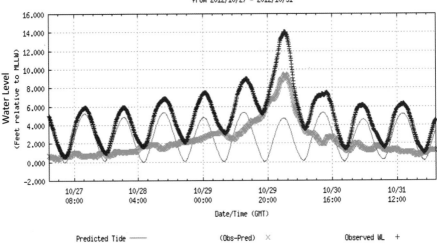

Predicted Tide ——   (Obs-Pred) ✕   Observed WL +

Figure 6.12 Water elevation at The Battery in lower Manhattan during Hurricane Sandy, October, 2012, with the blue line representing the predicted tide, the red line representing the measured water elevation, and the green line representing the difference between the expected (predicted) water elevation and the measured water elevation (NOAA, 2012).

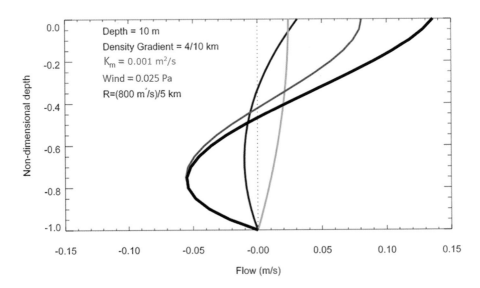

Figure 7.4 The vertical distribution of all the components of the estuarine circulation. The density gradient contribution is in orange; the wind is in red; the river flow is in green. The net flow is in blue, surface outflow and bottom inflow.

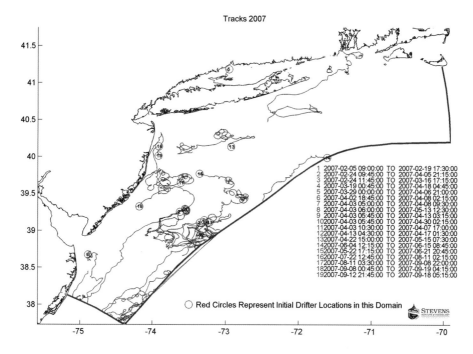

Figure 7.6 Currents deduced from satellite tracked drifters off the coasts of New York and New Jersey. The red line is the shelf break where water depths drop rapidly from 100 m or so to 1000 m.

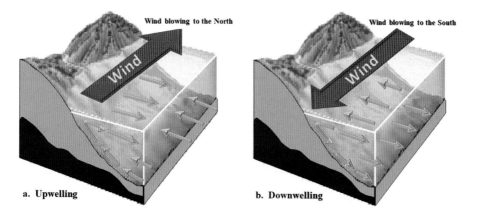

Figure 7.9 Coastal upwelling in the northern hemisphere with (a) wind blowing to the north and coastal downwelling with (b) wind blowing to the south. In the southern hemisphere, the relationship between the wind and currents is opposite to that shown. (Courtesy of Victor Rodriguez and Firas Saleh).

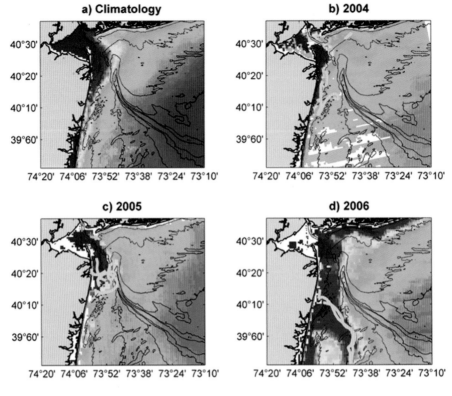

Figure 7.12 Moderate Resolution Imaging Spectroradiometer (MODIS) chlorophyll a images of the LaTTE study area for (a) April climatology (2004–2008), (b) 5 May 2004, (c) 4 April 2005, and (d) 28 April 2006. Higher chlorophyll concentrations (in red) are indicative of the presence of the Hudson River plume. Drifter deployments during the LaTTE experiments are shown in gray (Hunter et al., 2012).

Figure 7.13 Satellite infrared image of the waters off the East Coast of the United States showing the Gulf Stream, its meandering nature and its eddies.

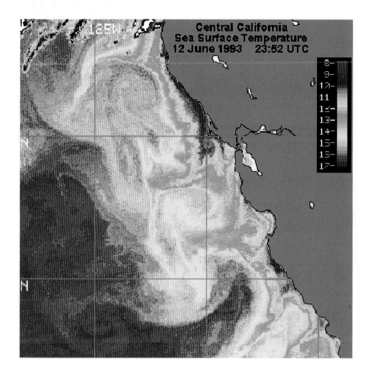

Figure 7.14 Satellite infrared image of the waters off the west coast of the United States taken June 12, 1983. The spacing of the horizontal lines is about 100 km. The upwelled coldest water is seen along the coast (Flament and Armi, 1985).

Figure 8.3 A view of an urban boundary in Beirut, Lebanon (www.trekearth.com/gallery/Middle_East/Lebanon/West/Beyrouth/Beirut/photo99978.htm).

Figure 9.10 Photo of Northern New Jersey beach with Sandy Hook in the background, ca. 1995 (photo by M. Bruno).

Figure 9.11 Accelerated beach erosion in front of a seawall in Long Island, NY, 1997 (photos by M. Bruno).

Figure 9.12 Aftermath of Hurricane Sandy at Mantoloking, New Jersey, October, 2012 (photo courtesy of US Army Corps of Engineers).

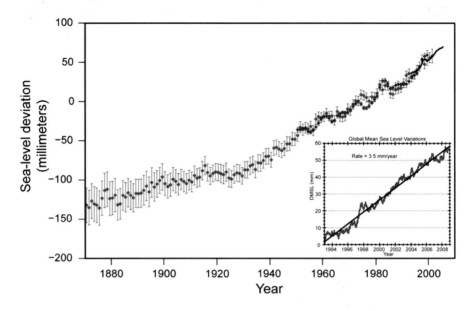

Figure 9.13 Annual averages of global sea level. The red line indicates the trend in sea-level since 1870. The dark line in the primary figure illustrates the measurements based on tide gauge data, with the last portion based also on satellite observations. The inset shows the sea level rise measured since 1993, during the period in which satellite measurements were available (IPCC, 2007).

Figure 11.4 The March 11, 2011 Tōhoku tsunami overtopping a seawall (courtesy of City of Miyako government).

Figure 11.3 Photo collage of damage in New York City from Hurricane Sandy (A. Seebode, US Army Corps of Engineers).

Figure 11.5 Aerial view of damage to Sukuiso, Japan, 1 week after the March 11, 2011 earthquake and tsunami (courtesy US Government, Dylan McCord, US Navy).

Figure 11.6 The Maeslant barrier (courtesy the Delft Institute for Port Innovation Research & Education).

Figure 11.7 Photo of a stone seawall (courtesy of Stevens Institute of Technology).

Figure 11.8 Photo of a stone groin field (courtesy of Stevens Institute of Technology).

Figure 11.9 Photos of a section of New Jersey shoreline following a beach restoration project (left) and in the aftermath of Hurricane Sandy (right) (courtesy of Stevens Institute of Technology).

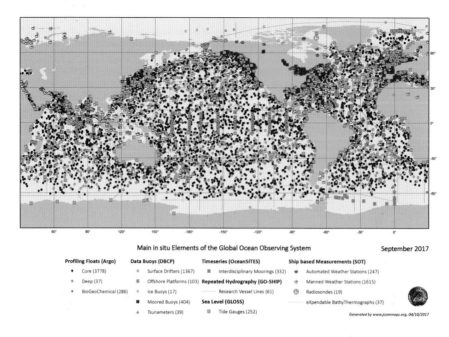

Main in situ Elements of the Global Ocean Observing System   September 2017

Profiling Floats (Argo)	Data Buoys (DBCP)	Timeseries (OceanSITES)	Ship based Measurements (SOT)
Core (3778)	Surface Drifters (1367)	Interdisciplinary Moorings (332)	Automated Weather Stations (247)
Deep (37)	Offshore Platforms (103)	**Repeated Hydrography (GO-SHIP)**	Manned Weather Stations (1615)
BioGeoChemical (286)	Ice Buoys (17)	Research Vessel Lines (61)	Radiosondes (19)
	Moored Buoys (404)	**Sea Level (GLOSS)**	eXpendable BathyThermographs (37)
	Tsunameters (39)	Tide Gauges (252)	

Generated by www.jcommops.org, 04/10/2017

Figure 12.1 September, 2017 status of the Global Ocean Observing System (World Meteorological Organization, 2017).

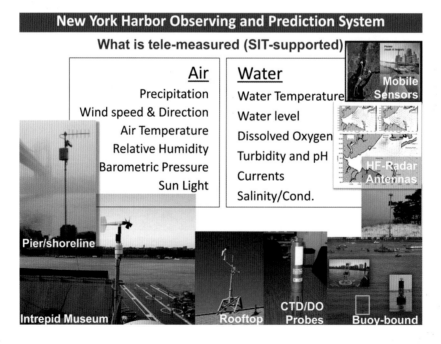

Figure 12.3 A photo mosaic of the measurement system types and locations employed in the NYHOPS ocean and weather measurement system (courtesy Stevens Institute of Technology).

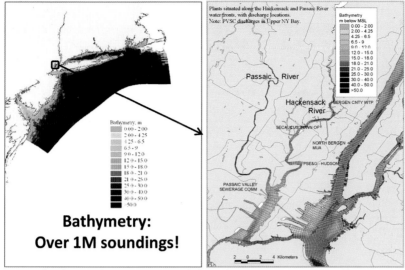

Bathymetry:
Over 1M soundings!

One high-resolution 3D model grid:150,680 water cells.
From 7.5km to <50m variable resolution. 7 coastal states.

Figure 12.4 NYHOPS computer model domain and high-resolution grid (courtesy Stevens Institute of Technology).

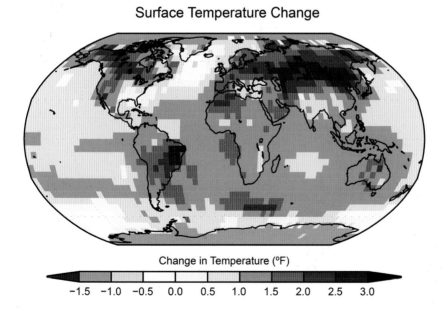

Figure 13.2 Surface temperature change (in °F) for the period 1986–2015 relative to 1901–1960 (USGCRP, 2017).

Rapid Emissions Reductions (RCP 2.6)

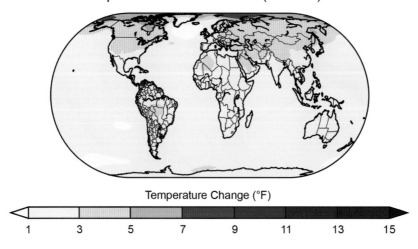

Temperature Change (°F)

1   3   5   7   9   11   13   15

Figure 13.6 The distribution of warming predicted by the lowest emissions scenario, RCP 2.6.

Continued Emissions Increase (RCP 8.5)

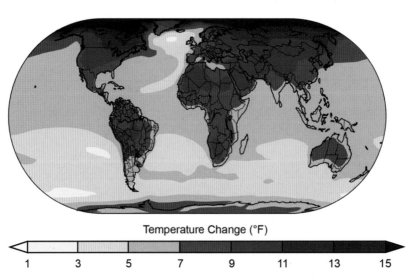

Temperature Change (°F)

1   3   5   7   9   11   13   15

Figure 13.7 The distribution of warming predicted by the highest emissions scenario, RCP 8.5.

11

Coastal Extreme Events: The Risks and the Responses

11.1   Introduction

In this chapter, we will examine the risks and impacts of coastal extreme events along densely populated coastlines. The continuing increase in population density along coastal regions worldwide in and of itself places more and more population centers at significant risk of loss of life, property, and the socio-technical-economic functions associated with the critical infrastructure on which modern society relies. In other words, the *consequence* component of the risk equation has increased steadily as the population distribution has continued to skew further toward urbanized coastal environments. With respect to the *probability* component of the risk equation, we are here focused on the natural hazards associated with the coastal ocean, namely extreme weather events produced by tropical and extratropical cyclonic storms, and tsunamis. As we have already discussed, this component has also increased as a consequence of climate change and human development along the coast. Warmer waters and atmosphere appear to be causing an increase in extreme storm frequency and intensity (IPCC, 2013), thereby increasing the probability of occurrence of damaging coastal storms. Perhaps more impactful and certainly more visible is the contribution to the probability of damage due to flooding associated with sea level rise. Simply stated, the water is getting closer and hence the probability of damage due to flooding is increasing. In most of the developed coastal regions on Earth, the position of the sea level relative to the land is increasing due to the combined effects of absolute sea level rise associated with climate change, and land subsidence associated with human activities such as coastal land development and the removal of underground water and oil. We should emphasize here that this is not a future problem; it is a here-and-now problem. Cities such as Norfolk,

Virginia are presently experiencing routine flooding during periods of high astronomical tide, with streets and low-lying buildings and infrastructure subject to inundation as frequently as twice per month during Spring Tide events. In a very real sense, this situation increases the probability of an extreme hazard occurring, as coastal storms that once produced only moderate flooding will now produce more severe and dangerous flooding. It also increases the consequence of extreme flooding events by subjecting a larger region of the coast to inundation and increasing the water level relative to the coastal built environment during extreme flooding events.

Our focus on flooding from the outset is quite intentional. Among all natural disasters worldwide, flooding is the primary cause of loss of life and property. And nowhere is flooding more impactful, more dramatic, and more frightening than along populated coastal areas, where the combination of geography and human investment produce the risk of catastrophic damage. Urban coastlines are almost by definition low lying and in close proximity to flood waters, often from multiple directions, e.g., the ocean, rivers, harbors, bays, and lakes. The consequence of flood events in these regions is often magnified by investment, not only in terms of the built environment (e.g., homes, roads, ports, and other infrastructure), but also the vulnerable and fragile natural coastal environment. Extreme coastal flooding events, most frequently associated with storm surges and tsunamis, can cause more changes to coastal landforms and ecosystems in a matter of hours than the changes associated with the "normal" influences, e.g., of tides, waves, and winds over many years.

We should remember that many, if not most, of the populated coastal regions of the world have been hosts to human populations for millennia. They are therefore valuable and worthy of protection and preservation not only because of their economic and environmental significance, but also because of their cultural and historical significance. This fact is unfortunately often ignored during policy debates surrounding the relocation of homes and communities from the coastal boundary, and population migration from low-lying coastal, delta, and island regions. Retreat in many of these communities is not a welcome option. As we will see later in this chapter, engineering solutions such as flood control systems and shoreline restoration can be employed in some situations with success. But these solutions are prohibitively expensive for many coastal communities, particularly those in emerging economies. They can also produce adverse impacts to the natural landscape and ecosystems. In a later chapter, we will describe a new approach to this dilemma that seeks to combine some elements of engineered solutions with policy and planning approaches that together create a coastal community that is by design more resilient to the inevitable coastal extreme event.

For the purposes of introducing this topic in more stark terms, and in order to identify particular lessons learned, we will describe the experiences of two communities – New York City during Hurricane Sandy, and the Oshika Peninsula of Tōhoku, Japan during the 2011 Tōhoku earthquake and tsunami. We have already spoken about storm surges, tsunamis, and other coastal events. Here we aim to place these phenomena into *context*, the context of populated coastal areas.

11.2   Extreme Events in Context

11.2.1   *Hurricane Sandy*

Excerpt from the Associated Press, November 3, 2012, Staten Island, New York, a neighborhood where 23 people died during Hurricane Sandy:

> *Most of the deaths were clustered in beachfront neighborhoods exposed to the Atlantic Ocean along the island's southeastern shore, an area of cinderblock bungalows and condominiums. Many of these homes were built decades ago — originally as summer cottages — and were not constructed to withstand the power of a major storm. Diane Fieros wept as she recalled how she and her family survived by huddling on the third floor of their home across the street from the ocean, watching as the waves slammed into the house and the water rose higher and higher, shooting through cracks in the floor. A few blocks away, several people drowned. "The deck was moving, the house was moving," she said. "We thought we were going to die. We prayed. We all prayed." Fieros rode out the storm with her two sons, her parents and other extended family members. She pointed to a black line on the house that marked where the water rose: at least 12 feet above the ground. "I told them, 'We die, we die together,'" she said, her voice cracking. "You saw the waves coming. Oh my God."*

Hurricane Sandy was a Category 1 hurricane prior to making landfall on October 29, 2012 along the coast of New Jersey, just south of New York City (see Figure 11.1). The storm's trajectory, nearly perpendicular to the coast with the strongest winds directed onshore, and its very low atmospheric pressure (940 mb), combined to produce a record storm surge along much of this region of the northeastern United States. The peak of the storm near New York City occurred at the time of high astronomical tide, further exacerbating the level of flooding and surprising residents and emergency management personnel with the rate at which the water level rose.

In order to place this event into historical context, we illustrate in Figure 11.2 the highest observed water levels since the late 1700s at The Battery in New York City, located at the southernmost tip of the island of Manhattan. The water

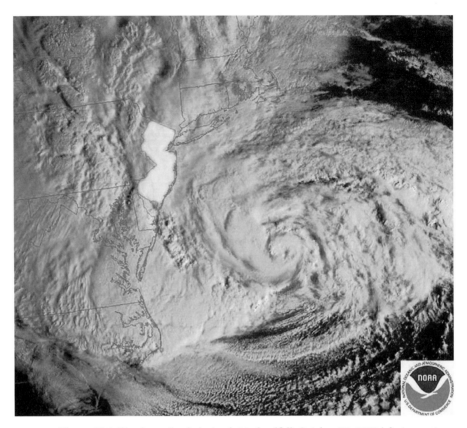

Figure 11.1 Hurricane Sandy just prior to landfall, October 29, 2012 (photo courtesy of NOAA).

elevations shown each represent the height above the mean sea level at the time of the measurement. The estimated water elevations during the 1788 and 1821 events are illustrated with 95% confidence bars. Note that the maximum water level during Hurricane Sandy was the highest observed over this 225-year period of time.

Figure 11.3 illustrates the scope of damage in New York City as a result of Hurricane Sandy. Although historic by virtually any measure, Hurricane Sandy was not among the most powerful hurricanes to strike the United States (that distinction goes to the Galveston Hurricane of 1900 as the deadliest, which caused up to 12,000 deaths; or Hurricane Katrina of 2005 as the costliest at US $108 billion [in 2013 dollars]; or Hurricane Camille of 1969 as the highest wind speed at 190 miles per hour when it struck the Mississippi coast). But as we have already discussed, risk can be calculated as the product of probability and consequence. Although Hurricane Sandy was clearly a low-probability event, the consequence of this storm striking one of the major metropolitan centers in

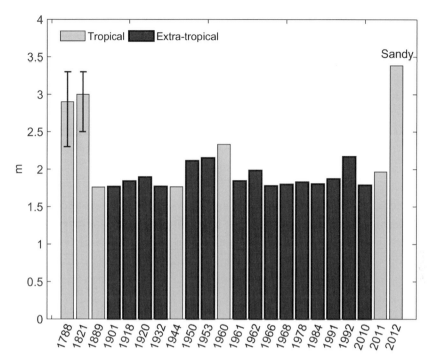

Figure 11.2 Water level in meters at Battery Park, NY during extreme storm events, including both tropical (hurricanes) and extratropical ("Northeasters") storms. The levels indicated for the 1788 and 1821 events have 95% confidence bars (courtesy of Philip Orton, Stevens Institute of Technology).

Figure 11.3 Photo collage of damage in New York City from Hurricane Sandy (A. Seebode, US Army Corps of Engineers). (A black-and-white version of this figure appears in some formats. For the color version, please refer to the plate section.)

the world was staggering. In immediate dollar terms, the storm ranked second in US history to Hurricane Katrina, with a total of US$71.4 billion in damages (in 2013 dollars). However, the impact of the storm persists to this day, both in terms of ongoing costs (e.g., persistent damage to subway and train tunnels due to saltwater intrusion into concrete walls) and in terms of ongoing coastal restoration and community rebuilding projects throughout the region.

The lessons learned during the storm were numerous and are still being examined today. We here summarize several of the key lessons, most if not all of which can be transferred to other coastal urban communities.

- Preparedness and response is an inexact science. The preparation for the storm's arrival took many forms, depending on the hazard type that was deemed most threatening. As an example, many of the facilities of the Port of New York and New Jersey prepared for the impact of high winds and placed critical equipment along sheltered waterfront areas. When the storm hit, the most impactful hazard was flooding, and as a result, the port facilities were submerged under several feet of ocean water and most of the critical equipment was lost. By contrast, the operators of the New York City subway system, the Metropolitan Transportation Authority (MTA), prepared for flooding by moving the subway cars to elevated tracks across the city. None of the cars were damaged, but there was recognition following the storm that only a slightly higher windspeed would have resulted in hundreds of subway cars possibly being tossed from the elevated tracks and onto the city roads and properties below.

- The critical lifeline infrastructure on which modern metropolitan areas rely (power, communication, transportation, healthcare, food and water) is complex, interdependent, and interconnected. This characteristic produces cascades of failure during a major disruption such as Hurricane Sandy. Power outages produced the urgent need for generators. But fuel stations could not supply fuel for the generators because they lacked the power to pump the fuel. Truck deliveries of fuel were severely hampered by safety regulations at the bridges and tunnels that connect Manhattan Island to the mainland. Fuel deliveries by barge were initially impossible because of the damage to waterfront facilities and docks. Critical life support systems such as hospitals had generators and fuel, but building practices and codes resulted in the generators being located in basements and first floors, exposing them to flooding and putting thousands of lives at risk. Communication – perhaps the most urgent need in the aftermath of an emergency – was severely limited by the fact that most of the population relies entirely on mobile phones, and there was no power to charge the phones.

- Planning and preparation for extreme events requires the proper translation of mostly scientific information – including uncertainty – to the public and decision-makers. This is a major challenge for coastal scientists and engineers, and will require the development of, among other tools, more sophisticated visualization capabilities that will allow decision-makers and emergency response teams (and the general public) to immerse themselves into the forecasted (or past) event and better understand the impacts and the necessary preparations and response. In the same way, warnings about the hazards associated with a particular event, whether they be wind or precipitation or flooding due to storm surge, should be cast in a fashion that points to specific action(s) that a community can take in order to protect lives and property.
- Ultimately, it is all about people. The New York Subway system was back in operation only days after the storm made landfall, in contrast to other transit systems in the region that required months. A significant reason was that the MTA went to great lengths to ensure that workers remained nearby during the days leading up to the storm and in the days immediately following. Problems were solved by people with years of knowledge about the system, often in real-time as the storm raged overhead.

In the year following the storm, the Federal government launched an international competition to consider innovative approaches to rebuilding in the New York metropolitan area with the aim of mitigating against future losses while also improving quality of life. This initiative, entitled Rebuild by Design, awarded nearly $1 billion to various projects across the region. But perhaps most significantly, it evolved into a significant international organization that is assisting communities around the world to become more resilient to disruption through research and design innovations (see: www.rebuildbydesign.org).

11.2.2   *Tōhoku Earthquake and Tsunami*

The magnitude 9.0 Honshu, Japan earthquake (38.297° N, 142.372° E, depth 30 km) on March 11, 2011, generated a tsunami that caused 15,890 deaths, 2,590 missing and presumed deaths, and 6,152 injuries in 12 Japanese prefectures (NOAA, 2015). In addition to this staggering human toll, the earthquake and tsunami combined to cause damages totaling US$220 billion in Japan, making this the most costly natural disaster in recorded history. The earthquake was the most powerful to strike Japan since an instrumented measurement

Figure 11.4 The March 11, 2011 Tōhoku tsunami overtopping a seawall (courtesy of City of Miyako government). (A black-and-white version of this figure appears in some formats. For the color version, please refer to the plate section.)

program began in 1900. The tsunami had a maximum wave run-up height of 126 feet (38.9 m), with the water overtopping even the highest of coastal seawalls, as shown in Figure 11.4. The flooding caused a nuclear disaster at the Fukushima I (Daiichi) Nuclear Power station.

Figure 11.5 illustrates the utter devastation along the coastline caused by the tsunami. Note that nearly all of the structures in this low-lying area of Sukuiso were completely removed by the force of the series of tsunami waves.

As was the case with Hurricane Sandy, the Tōhoku tsunami has had a lasting influence on the assessment of, and response to, coastal natural hazards worldwide. The Sendai Framework was adopted by UN Member States on March 18, 2015 at the Third UN World Conference on Disaster Risk Reduction in Sendai City, Japan. As will be described more fully in a later chapter, the Sendai Framework is a 15-year, voluntary, non-binding agreement that recognizes that each nation has the primary role to reduce disaster risk but that responsibility should be shared with local government, the private sector, and other stakeholders. It aims for the following outcome: "The substantial reduction of disaster risk and losses in lives, livelihoods and health and in the economic, physical, social, cultural and environmental assets of persons, businesses, communities and countries."

Of course, Japan lies in one of the most active seismic regions on Earth, and as an island nation, retreat from the coastline is not a viable option for most if not

Figure 11.5 Aerial view of damage to Sukuiso, Japan, 1 week after the March 11, 2011 earthquake and tsunami (courtesy US Government, Dylan McCord, US Navy). (A black-and-white version of this figure appears in some formats. For the color version, please refer to the plate section.)

all of the nation's urban areas. Elevation of existing built environments (a strategy employed in Galveston, Texas after the 1900 storm surge) and/or the construction of still higher coastal seawalls would seem to be the only path toward a reduction of the risk of disaster should another tsunami of equal or greater magnitude strike the coast.

An important lesson-learned that emerged from this catastrophic tsunami and the crisis at the Fukushima nuclear power plant was that – as was found during Hurricane Sandy – people are the primary source of response, recovery, and resiliency. The now legendary courage and skill of the small group of nuclear workers who volunteered to remain behind and perform a safe shutdown of the reactors likely saved countless lives. It would seem that no amount of training or preparation could have prepared them for the moment when they needed to make this decision. But the fact that trained professionals were on site at the moment of crisis was, as in the case of Hurricane Sandy, an essential first ingredient to producing a resilient system and a recoverable event for the community.

The catastrophic impact to the Fukushima nuclear power plant also forced a global conversation about the siting of critical infrastructure in hazard-prone locations, even when the hazards are very low probability (recall that risk is a strong function of the consequence of an event). Rare but very consequential events must be considered when performing a risk assessment along a densely populated coastal region. Although challenging from a political and public policy standpoint, such conversations must be conducted in an open fashion, and in terms that decision-makers and the general public can understand. The imperative for communication, and for science-informed policy and decision-making, has never been greater, especially given the combined phenomena of population migration to coastal cities and climate change.

11.2.3   *The Black Swan*

It has been popular in some circles to explain the all-too-frequent lack of preparation or adaptation to coastal hazards as a result of the unexpected nature of the event, be it a major hurricane making a landfall near New York City or a tsunami striking a coastal nuclear power plant. Recently, such events have been referred to as "Back Swan" events, defined by Taleb (2010) as events that are not predictable via conventional means but that in hindsight appear almost obvious. Clearly, as shown in Figure 11.2, New York City had experienced major flood events prior to Hurricane Sandy. And the coastline of Japan has experienced major earthquakes and tsunamis numerous times, as recently as the 1995 Kobe earthquake. But it often appears that whether it is for economic reasons or public policy concerns, communities seem unable to invest in major hazard mitigation projects until a major event has actually occurred. Perhaps the best one can hope for is that once a major event has occurred, the response includes the development of a forward-looking strategy that addresses not only the vulnerabilities observed during the event, but also the threat(s) associated with other, more impactful events. We will here examine a few well-known examples of such an approach, beginning with the "Delta Plan" in Holland.

11.3   Response to Extreme Threats

11.3.1   *The Delta Plan*

Holland is a nation famous for its tenacity and ingenuity in combating flooding along its low-lying coastal and riverside regions. Over centuries, the Dutch developed novel designs of flood mitigation structures, construction techniques, and land use management strategies that largely served the population well.

This all changed in 1953.

The North Sea Flood of 1953 caused catastrophic flooding in Holland and nearly 2,000 deaths. It also prompted the creation of a "Delta Works

Commission" tasked with examining the causes of the massive flooding with the aim of developing preventative measures to ensure that such flooding would never occur again. The result of the Commission's work was the "Delta-plan," a complex set of construction projects including storm surge barriers, inland dams, dikes, and levees. The complexity of this undertaking was magnified immensely by the seemingly conflicting need to ensure that the ocean access to the critical ports of Rotterdam and Antwerp remained in place. The Commission and elected officials recognized at an early stage that success would require a greater understanding of ocean and riverine dynamics in the region, as well as advances in structural engineering, materials engineering, construction management, and other areas of science, engineering, and land use planning.

There followed significant government investment in scientific research programs to advance knowledge in hydraulics, weather and ocean dynamics, engineering, and resource management. Over time, the nation became an exporter of knowledge and design in the broad area of flood protection and coastal engineering. This home-grown capacity for research and design has led also to a community that is perhaps the most aware of any in the world, not only of the risks of residing in a flood-prone region, but also the potential to reduce these risks through innovative engineering design and land use management, and the need to remain vigilant as conditions change.

In the final stages of the implementation of the plan, two extraordinary storm barriers were constructed in the 1990s, the "Hartelkering," completed in 1997 in South Holland, and the Maeslantkering, also completed in 1997 near Rotterdam. The barriers remain open to allow navigation at all times and are closed only in advance of a forecasted extreme storm surge, or when the barrier is closed for testing to ensure reliability. Figure 11.6 illustrates the Maeslant barrier, as it is known in English.

The "Delta Project," as it is now known, has continued to this day, with updates and maintenance performed as warranted either as a result of inspections or as a result of a re-assessment of risk. The latter factor is a primary point to be made here in our discussion. The Delta Law, and the later (2009) Water Law, requires that the risk of a flooding event be held to no more than 1 in 10,000 years along the most populous areas of the coast. As one might expect, the risk of such an event changes when new scientific evidence emerges of a past event that exceeds previously known events, or when an extreme event actually occurs along the Dutch coast or perhaps even elsewhere in the North Sea that alters the extreme event statistics. The Dutch government has recently included an assessment of the long-term influence of sea level rise as a required component of the assessment of risk of future extreme flooding events.

Figure 11.6 The Maeslant barrier (courtesy the Delft Institute for Port Innovation Research & Education). (A black-and-white version of this figure appears in some formats. For the color version, please refer to the plate section.)

11.3.2   The Thames Flood Barrier

The 1953 North Sea flood, which prompted the creation of the Delta Plan in Holland, also prompted calls in London for a strategy to protect the city from flooding due to storm surges that would at that time be free to propagate up the Thames River and impact the city and points beyond. Construction on what became known as the Thames Barrier began in 1974 and was completed by 1982. The Thames Barrier is composed of 10 individual steel gates, each of which lie on the river bed, only to be raised into a vertical position during periods when flooding threatens the city.

At this point the reader might wonder exactly how often a flood gate system such as the Thames Barrier or the Maeslant Barrier has needed to be closed to protect people and property. As of October, 2017, the Thames Barrier has been closed 179 times since it became operational in 1982. Of these closures, 92 were in response to the threat posed by extreme tidal (ocean) flooding and 87 were to protect against combined tidal/river flooding (www.gov.uk/guidance/the-thames-barrier).

11.3.3   Venice

For centuries one of the world's most important port cities, Venice, Italy, has remained an iconic urban center steeped in history, architecture, music, and art. Composed of numerous islands connected by bridges and canals, Venice has since its earliest days also been prone to flooding from storm surges and extreme rainfall events. Over the last century, both the frequency and magnitude of flooding have increased dramatically as a result of the combined influences of sea level rise and land subsidence due to human activities. The floods have threatened historic buildings, adversely impacted the vital tourism industry, and severely impacted the residents' quality of life.

In the late 1980s the city began the planning and design effort to create a flood protection system to guard against extreme storm surges from the Adriatic Sea. The project is known as MOSE (after the Italian **Mo**dulo **S**perimentale **E**lettro-meccanico, or Experimental Electromechanical Module). Expected to be operational sometime in the period between 2018 and 2020, the MOSE project consists primarily of mobile gates designed to be raised from the seafloor upon notification that the real-time water elevation sensors and/or computer predictive models indicate that a threatening flood event will soon occur (see www.mosevenezia.eu/?lang=en). The gates will function independently at the three entrances to the Venice Lagoon. The use of independent storm barriers at the three Lagoon inlets provides for flexibility in the operation of the system, allowing, for example, the raising of the gates at one or two inlets if the anticipated flooding is due to a severe wind event and therefore localized in only one zone of the protective system. It should be mentioned that in this region of the coast – as in many if not most coastal ocean regions – storm surge events are often accompanied by extreme rainfall. Time will tell whether the city will eventually need to turn its attention to controlling the flood waters associated with the drainage of land areas within and landward of Venice.

11.4   Response to More Common Coastal Hazards

The previous section discussed a few examples of the response by densely populated coastal communities to the threat posed by extreme flood events from storms. But what about the threat posed by more frequent coastal events? Certainly, it is not only the 200-year or even 100-year event that threatens communities and supporting infrastructure located along the coast. In fact, let us recall our discussion in Chapter 9 regarding the potential for significant annual fluctuations in the position of a shoreline (and hence the threat to coastal structures), and the possibility of very rapid beach erosion in front of coastal structures even in the absence of a single extreme storm event, as shown in Figures 9.1 and 9.11, respectively. Clearly, urban coastal communities must be cognizant of the "everyday" threats to homes and infrastructure posed by proximity to the dynamic coastal environment, where wind, waves, and shifting sands can produce unexpected or even threatening conditions without warning. But what is the best strategy for dealing with these threats?

11.4.1   *Traditional Approaches to Coastal Hazards*

So-called "traditional" approaches to coastal protection include "hard" strategies that center on the construction of structures, and "soft" strategies that involve the creation or restoration of a beach seaward of coastal structures as a buffer to elevated ocean tides and waves. The choice of strategy often depends on the local policies and regulations, in particular regarding the placement of structures along the coast. However, other considerations including, e.g., access to the beach, aesthetics, maintenance, and liability issues, will often dominate the decision-making. All else being equal, the process of narrowing down the approaches to be considered can be summarized as a series of decision points:

- Detailed analysis of the ocean and beach conditions at the site:
    - wave height, period, direction, including average, seasonal, and maximum
    - water elevation, including astronomical tides and storm surges
    - beach width and height
    - sand (sediment) characteristics and transport (magnitude and direction)
    - wind
- Risk assessment to identify the most likely hazards to property and infrastructure:
    - flooding/breaching?
    - wave action and/or overtopping (wave run-up over a beach or structure)?
    - beach erosion and/or scouring at the base of buildings and infrastructure?
    - consequence of failure – property damage, property loss, loss of life?
- Availability of required materials for the preferred solution, e.g., stone or concrete for a seawall, or sand for beach restoration.

11.4.1.1 Structural Approaches

The construction of coastal protection structures surely dates back to the earliest human settlements along the ocean. Although the construction materials, and the form and function of these structures has changed across the centuries, they all share a common aim – to guard against damage to property and infrastructure caused by:

- Coastal flooding from the open ocean, or from landward bays, lagoons, lakes
- Wave attack
- Beach erosion and scouring, which can undermine coastal structures

Coastal protective structures fall into the following categories:

1. Shore-parallel structures
   - Seawalls
     Seawalls are structures built along a shorefront area with the aim of protecting a landward area against flooding and/or wave action from the ocean side. These structures are typically constructed of either concrete or stone, to a height that is based on a determination of the probability of occurrence of specific water level elevations and wave heights, coupled with an assessment of the "level of acceptable risk". That is to say, the design height of a seawall is based both on the expected future environmental conditions and on the community's tolerance for risk of property damage, property loss, and – in extreme events – threat to human lives. Figure 11.7 illustrates an example of a stone seawall.

2. Shore-perpendicular structures
   - Groins
     Groins (also spelled "groynes") are structures placed across a beach with the aim of trapping sand that moves naturally along the shoreline (see Chapter 9) so that the beach remains wider and perhaps higher and thereby provides defense against wave action and flooding. These structures are typically constructed of either stone or wood, although in recent years there have been groin structures constructed using very large sand-filled geotextile tubes. Figure 11.8 provides an illustration of a few large stone groins. Note that a rather large groin "field" is also depicted in Figure 9.10.

11.4.1.2 Non-Structural Approaches

The use of non-structural approaches in the mitigation of coastal hazards along developed coastlines can, in some cases, be attributed to the absence or cost of the construction materials required for most structural approaches. However, more often the non-structural approach is prompted by concerns about the potential for adverse impacts from the presence of structures on or near the beach. For example, since groins are by definition installed for the

Figure 11.7 Photo of a stone seawall (courtesy of Stevens Institute of Technology). (A black-and-white version of this figure appears in some formats. For the color version, please refer to the plate section.)

purpose of trapping and retaining alongshore-moving sediment, it stands to reason that beach areas "downstream" from the groin will be deprived of sediment that would otherwise flow to them. Although this impact is somewhat visible in Figure 11.8, it is much more evident in Figure 9.10. This approach clearly has the effect of choosing "winners" and "losers" in terms of which shorefront area will enjoy a stable beach. Likewise, the presence of a seawall can have a negative impact on the beach, in particular if the seawall is impacted by ocean wave action. This can happen if and when an extreme storm event causes a storm surge large enough to bring the wave action up to the seawall, or if beach erosion is allowed to continue to the point where the seawall experiences wave attack during even normal high tides. In these cases, the refection of even a small portion of the wave energy at the base of the seawall will cause scouring of the sand in front of the structure and can lead to undermining and structural failure. Wave reflection can also have detrimental impacts to adjacent shoreline areas, if the reflected wave energy is additive to the incoming wave energy, resulting in increased wave heights and potentially higher rates of beach erosion.

For all of these reasons, a non-structural approach that has been used with success worldwide is the restoration or enlargement of the sand beach fronting the shoreline. This approach is often referred to as "beach nourishment". The

Figure 11.8 Photo of a stone groin field (courtesy of Stevens Institute of Technology). (A black-and-white version of this figure appears in some formats. For the color version, please refer to the plate section.)

process involves the placement of sand in sufficient volume and at the desired height and offshore slope so that the new beach provides the desired level of protection to the community. Of course, the additional benefit of this approach is that the community gains an expanded beachfront for use by residents and in many cases, tourists. There are a few key considerations when designing and constructing beach restoration projects:

- Method of placement. Before the advent of safe and effective offshore sand dredging, most beach nourishment projects employed trucks to deliver sand to the shorefront. This made the process very costly and disruptive. Most such projects, except for very small projects, are now accomplished using ocean-going dredges that bring sand up from the ocean floor and pump the sand up and onto the beach.
- Origin and type of sand. Experience has demonstrated that the most successful beach nourishment projects are those that employ sand that is identical or close in characteristics (sand grain size and material) to the native, natural beach sand at the project site. This appears to be doubly important in areas where the beach is an important component of a vibrant ecosystem, e.g., turtle nesting areas or shellfish areas.
- Time of placement. In most cases, it can take several months for a newly placed beach to adjust to the natural forces (primarily waves) that shape the beach. For this reason, and especially in areas where there is a strong seasonal variability in the wave climate, there needs to be careful consideration of sand placement during an expected extended period of relatively low-energy wave activity.

Figure 11.9 illustrates the beach along a section of the New Jersey coastline just after a beach nourishment project (on the left), and soon after Hurricane

Figure 11.9 Photos of a section of New Jersey shoreline following a beach restoration project (left) and in the aftermath of Hurricane Sandy (right) (courtesy of Stevens Institute of Technology). (A black-and-white version of this figure appears in some formats. For the color version, please refer to the plate section.)

Sandy (on the right), with the common location indicated with a red arrow. Note the "bulge" in the shoreline just after the beach restoration project was completed. This is quite common immediately after these projects, and prior to the beach adjusting to the waves and currents (alongshore and onshore-offshore) in the weeks and months that follow. The shoreline receded during the storm as a result of the extremely high storm surge and accompanying wave action, but the beach remained sufficiently wide to provide protection to the small town just landward. Many other such towns experienced near-total or total destruction during the storm.

11.4.2   Non-Traditional Approaches to Coastal Hazards

Over the last 10–20 years, there has been a gradual movement of coastal engineers, planners, and natural resource managers toward the adoption of new approaches to the mitigation of coastal hazards along developed shorelines. A primary feature of most of these new approaches has been an effort to:

- Understand the root cause of the perceived or calculated risk of loss of life and/or property. These may include, for example, erosion of the shorefront due to the loss of sediment supply. Or the loss of natural protective features such as dunes, wetlands, offshore islands, or coral reefs.

- Understand the natural processes that provide protection to coastal landforms, including beach erosion, which causes a reduction in beach width but also causes a reduction in wave energy.
- Develop coastal protection strategies that can mimic nature to the extent possible and/or that can replace the functioning of pre-existing natural protective features.

Examples of these new approaches include the large-scale restoration of off-shore barrier islands and coastal wetlands along the Gulf Coast of Louisiana in the years following Hurricane Katrina. This effort was prompted by the recognition that decades of construction of dams and levees along the Mississippi River and its tributaries had severely diminished the amount of sediment being delivered to the Gulf of Mexico (see our discussion about the impact of Egypt's Aswan High Dam in Chapter 9), resulting in the slow but steady loss of wetlands and offshore islands that had long provided natural protection to the coast and to inland areas by dampening the wave action and the storm surges. On a smaller scale, many communities around the world are adopting the strategy of building (or rebuilding) vegetated dunes along the landward edge of beaches. These dunes provide very effective protection against both flooding and high-water wave attack. In a real sense, the dune is "sacrificial" in that the erosion of the sand and vegetation expends wave energy that would otherwise be expended on homes or infrastructure. The community must recognize of course that it is essential to restore the dune and its vegetation soon after a major storm event.

In cases where structures and infrastructure are threatened as a result of shoreline recession caused by the reduction or elimination of sediment supply to the coast, many communities have adopted a long-term strategy of essentially replacing the original sediment source with periodic beach nourishment, often with an established nourishment location (commonly referred to as a "feeder beach") and an established schedule of sand placement (usually between 5 and 10 years, depending on the rate of shoreline recession).

The use of man-made materials to restore or replace natural offshore reefs is not yet widespread, but advances in our understanding of the role of natural reefs in buffering wave action and storm surges, coupled with advances in materials science (including materials that can encourage natural coral colonization), give hope that such projects may become an important component of coastal hazard mitigation projects in areas of the world with coral reefs. One perhaps surprising example of artificial reef building is the current effort in New York Harbor to seed oyster beds in areas where oysters were plentiful up until the twenieth century (see https://billionoysterproject.org). These beds provide a living buffer to wave action and storm surge in the area, and the oysters themselves, because they are filter feeders, actually improve the water quality.

Recent strategies in flood-prone areas in Holland have included novel approaches such as floating homes designed to simply rise and fall during extreme riverine flooding events, and dikes that have been transformed into natural landscapes containing parks and bike paths.

11.5   Focus on the Community

There is slowly emerging a consensus view among engineers, architects, and planners that:

1. It is the preservation of the critical socio-technical-economic function that is essential to ensuring that a community is resilient against even the most unexpected of extreme events, rather than the actual physical infrastructure. This will be discussed more fully in a later chapter, but in the context of the present discussion it suggests that protective systems should be designed to "degrade gracefully" during an extreme event, so that the critical function that the system provides (e.g., power or communication) is not entirely lost, but can provide some minimally acceptable level of function until full recovery can be achieved.

2. A cohesive community is a more resilient community in the face of disaster. Recall our lessons learned about the irreplaceable role of humans in the preparation for and response to both Hurricane Sandy and the Tōhoku earthquake and tsunami. Imagine an entire community that is knowledgeable and well-prepared and possesses the strong empathy for others that is required to provide assistance in the midst of an extreme event. The design of mitigation measures must therefore enhance community cohesion and not detract from it. The use of natural landforms that double as protective features only when needed can produce an enhanced quality of life. The assurance of strong communication channels between decision-makers and the general public can inspire confidence. Among the challenges that exist with respect to this communication, as well as communication among the residents themselves, are the socio-economic and language barriers across the populations of so many of the world's coastal urban areas. In a very real sense, the resilience of the world's coastal communities will rely as much if not more on the strength of the social fabric of these communities as on the engineered systems that we create to protect them.

12

Coastal Ocean Observing Systems

12.1   Introduction

The notion of "coastal ocean observing systems" is actually a very recent addition to the lexicon of the global oceanography community. In fact, it is only over the last 20 or so years that we have been pursuing what we would term "observing". There is a fundamental difference between the traditional, hypothesis-driven ocean sampling that first dominated our attempts to understand ocean dynamics and the more recent approach of continuous ocean observations. As oceanography adjusted to the Information Age in the 1990s, so too did our approach to gaining a deeper understanding of the complex interactions among air, water, land, and human influences. In sharp contrast to Darwin's voyages on the Beagle in the early 1800s, today's ocean scientists have, from the dawn of the Internet and the advent of satellite-based measurement systems, been afforded the opportunity to simply "watch" as the ocean and its biogeochemical systems respond to natural and man-made influences.

The modern oceanographic community has watched with a careful eye advanced instrumentation and increasingly sophisticated analytical and visualization tools. We have learned of the existence of causal relationships among physical, chemical, and biological systems that we never knew existed. And we have done this during periods of time, such as during extreme weather and ocean events, when traditional ship-based sampling programs were simply not possible – but when many of the most impactful changes occur, as we have already learned.

Put simply, the coastal ocean is a very complex place and it is also a very difficult place in which to work. But modern communication and measurement technologies have changed our work forever. This chapter will discuss the various challenges associated with coastal ocean observations, while also pointing the way toward solution paths to support integrated observing and decision-making systems.

12.2   A Bit of History

In addition to the obvious role of advanced measurement and communication technologies, a key contributor to the development and sustainment of today's coastal ocean observation systems is the longstanding tradition of international cooperation and collaboration in the study of the world's oceans, which know no national boundaries and influence all of mankind. In the late 1990s the UN's Intergovernmental Oceanographic Commission (IOC), the World Meteorological Organization (WMO), the United Nations Environmental Program (UNEP), and the International Council for Science (ICSU) set the stage for the development of a coordinated, multi-national effort to design, deploy, and maintain a global system of ocean and weather observation platforms. This resulted in what is referred to today as the Global Ocean Observing System (GOOS). The system is a voluntary program with sensors and data management systems contributed by many nations and coordinated by the Intergovernmental Oceanographic Commission (IOC) of UNESCO. Figure 12.1 illustrates the status of the GOOS as of September, 2017, with the various sensor types indicated (see www.jcommops.org).

In Figure 12.1, SOT refers to the Joint Technical Commission for Oceanography and Marine Meteorology's (JCOMM's) Ship Observations Team, GO-SHIP refers to the Global Ocean Ship-based Hydrographic Investigation Program, GLOSS refers to the Global Sea Level Observing System, DBCP refers to the Data Buoy Cooperation Panel, and Argo is the name given to a global array of temperature/salinity profiling floats. For more information regarding these sensor systems, and the history and status of the GOOS, the reader is referred to www.goosocean.org.

In 2005, UNESCO published a report entitled "An Implementation Strategy for the Coastal Module of the Global Ocean Observing System" (UNESCO, 2005). This report built on the work up to that point by coastal ocean scientists worldwide, and it recognized the fact that as opposed to the deep ocean and climate focus of the GOOS, the various coastal ocean observation systems in place at that time around the world were (and still are) put in place to address place-specific issues such as water quality, beach erosion, safe navigation, and fishing, to name a few. The report advocated for the creation of a Global Coastal Network comprised of GOOS Regional Alliances. The report further advocated for the inclusion of sensors to provide a comprehensive set of variables including temperature, salinity, ocean currents, wave characteristics, sea level, shoreline

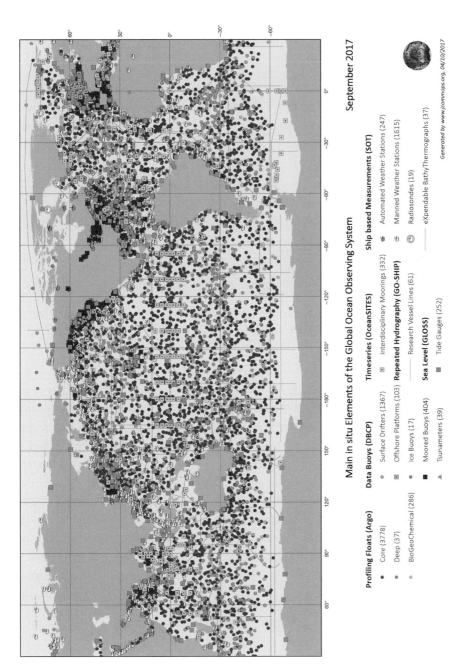

Main in situ Elements of the Global Ocean Observing System

September 2017

Profiling Floats (Argo)
- Core (3778)
- Deep (37)
- BioGeoChemical (286)

Data Buoys (DBCP)
- Surface Drifters (1367)
- Offshore Platforms (103)
- Ice Buoys (17)
- Moored Buoys (404)
- Tsunameters (39)

Timeseries (OceanSITES)
- Interdisciplinary Moorings (332)
- **Repeated Hydrography (GO-SHIP)**
  - Research Vessel Lines (61)

Sea Level (GLOSS)
- Tide Gauges (252)

Ship based Measurements (SOT)
- Automated Weather Stations (247)
- Manned Weather Stations (1615)
- Radiosondes (19)
- eXpendable BathyThermographs (37)

Generated by www.jcommops.org, 04/10/2017

Figure 12.1 September, 2017 status of the Global Ocean Observing System (World Meteorological Organization, 2017). (A black-and-white version of this figure appears in some formats. For the color version, please refer to the plate section.)

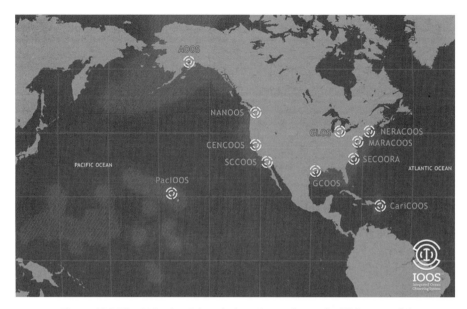

Figure 12.2 The 11 Regional Associations that make up the US Integrated Ocean Observing System (NOAA. 2017c).

position, bathymetry, sediment characteristics, dissolved oxygen, nutrients, biomass, water quality (fecal indicators), and optical properties.

In the United States, building on the progress achieved in building out several regional coastal observing systems around the nation, NOAA established in 2006 the Integrated Ocean Observing System (IOOS) Program Office. Federal funding soon followed and helped to establish a set of 11 Regional Associations. The locations of these systems are shown in Figure 12.2. Prior to their inclusion in the IOOS, the sensor systems were evaluated based on their contribution to the seven Societal Benefit Areas identified by the National Ocean Research Leadership Council:

- Detecting and forecasting oceanic components of climate variability
- Facilitating safe and efficient marine operations
- Ensuring national security
- Managing resources for sustainable use
- Preserving and restoring healthy marine ecosystems
- Mitigating natural hazards
- Ensuring public health

In addition to the obvious benefits of providing large-area ocean observations, coordinated efforts such as GOOS and IOOS also facilitate the sharing of best practices among organizations, and inspire the development of new technologies and new approaches to data gathering, analysis, and dissemination. Collaborations among universities, industry, and governments have resulted in significant advances in our ability to provide continuous, real-time observations to various user communities, most of it via the Internet. Examples of these coastal ocean observation systems from around the world include the following:

- Urban Ocean Observatory at Davidson Laboratory: http://hudson.dl .stevens-tech.edu/maritimeforecast/
- Pacific Islands Ocean Observing System (PacIOOS) Voyager: www .pacioos.hawaii.edu/voyager/
- Mediterranean Oceanography Network for the Global Ocean Observing System: www.mongoos.eu
- North West Shelf Operational Oceanographic System: http://noos .eurogoos.eu
- Japan Meteorological Agency: www.jma.go.jp/jma/indexe.html
- GOOS Brazil: www.mar.mil.br/secirm/ingles/goos.html
- Integrated Marine Observing System: http://imos.org.au/home/

12.3   Integrated Coastal Ocean Observing Systems

The ultimate aim of an integrated coastal ocean observing system such as IOOS is to measure the important coastal ocean variables and to convert those data into usable information. Worldwide, the evolution of these systems has often included the integration of three separate subsystems:

1. the measurement (monitoring) subsystem
2. the data analysis-modeling subsystem
3. the data management and communications subsystem

These integrated systems ideally provide ocean state estimates (past, present, and future) to a known degree of accuracy based on the integrated use of the real-time ocean observations, data assimilative model predictions, and cyber-infrastructure tools. When properly designed and implemented, the ocean state estimates should produce "actionable" information regarding physical, chemical, and biological characteristics delivered to all of the various user communities. Such information can range from scientific findings, to operational products (e.g., to support safe navigation), and products in support of public education and public policy, to name a few examples. Ideally, these ocean state estimates should also be incorporated into programs that quantify the drivers of change across time scales from episodic (e.g., storm events) to climatic along our coastal ocean regions. These drivers might include atmospheric forcings, land-based factors, and perhaps most important, human influences. In order to be successful and sustainable over the long term, integrated coastal ocean observation systems should bridge the gap between research and technology development, thereby integrating the activities of the R&D communities and the operational communities across academic, private, and government organizations. Let us here take a look at each of the identified types of observing subsystems.

12.3.1   The Measurement (Monitoring) Subsystem

As mentioned earlier, the initial discussions about a coastal ocean component of GOOS advocated for systems that could provide measurements of a comprehensive set of variables including temperature, salinity, ocean currents, wave characteristics, sea level, shoreline position, bathymetry, sediment characteristics, dissolved oxygen, nutrients, biomass, water quality, and optical properties. Clearly, and as indicated in the examples of coastal observation systems provided earlier, no one system can be expected to provide information regarding all of these variables. However, as sensing technologies have improved, we are fast approaching the time when we will have the capability to provide real-time, highly accurate measurements of these variables, and others. And as we move toward more and more strongly integrated systems across regional scales of 100s of miles, there is the opportunity to combine a diverse set of instrumentation and data products into a comprehensive view of the real-time, present state of the coastal ocean.

As an example of a measurement subsystem, The New York Harbor Observing and Prediction System, or NYHOPS (Georgas et al., 2009; viewable at www.stevens.edu/maritimeforecast) provides operations-critical information to the Port of New York and New Jersey and many other users. Real-time oceanographic information within New York Harbor is obtained using various sensors placed at strategic locations to monitor the current state of the estuarine environment. The challenge of providing this information is particularly acute in urban ocean environments, for several reasons:

1. Urban ocean regions are usually located in shallow coastal areas and estuaries where oceanographic and atmospheric conditions exhibit high spatial and temporal variability due to the influences of:
   - natural freshwater inflows, and anthropogenic freshwater inflows such as industrial and wastewater treatment plant outflows, and combined sewer overflows
   - the complex interaction of the tide with estuarine flows (already described elsewhere) as well as with human-altered flows in, e.g.,

dredged navigation channels and in and around dredged, heavily altered waterfront dock facilities
- micro-climate caused by, e.g., high rise waterfront structures that can block the wind and the sunlight
- bottom and land topography that is subject to constant change as a result of human activities including coastal development
2. Urban ocean regions are usually the home of an extraordinarily intense and complex array of human activities, including commercial transport characterized by ever-larger cargo vessels, ferry transport systems characterized by ever-faster ferry vessels, recreational vessels including power boats and sailboats, and human-powered vessels such as kayaks and paddleboards
3. Complex ocean and atmospheric conditions associated with both location and human activities, including high turbidity, strong stratification, strong ocean currents, and fog
4. And limited shoreline access as a result of the fact that urban ocean waterfronts are usually lined with private development

As a result of these complexities, urban ocean measurement systems must be made rugged and robust against both intentional and unintentional contact, including collisions. They must be designed to be dependable and low-maintenance, preferably with on-board power supply. And they must be strategically located in the region of interest in order to provide the type of measurements (e.g., water temperature and salinity, wind speed and direction) at locations that taken together can provide an accurate and relevant assessment of the state of the ocean and the weather. We highly recommend the design process utilized in the creation of NYHOPS, wherein a three-dimensional ocean simulation model was employed with limited initial ocean and weather observations in order to help identify the most optimum locations for sensor placement, from the standpoint of characterizing the ocean and weather conditions in the estuary.

The NYHOPS network sensors, all of which provide their data in real-time, consist of:

- Six shore-based salinity, temperature, turbidity, and water level sensors, and two additional water level sensors. These include conductivity-temperature-depth (CTDs) sensors, which record a point measurement of temperature and salinity, and optical backscatter sensors (OBSs), which measure relative turbidity, or the measure of the amount of suspended material in the water.
- Two moored platforms in the estuary containing near-surface and near-bottom salinity, temperature, turbidity, and water level sensors (CTDs and OBSs).

- Two Acoustic Doppler Current Profilers (ADCPs) mounted on navigation aids (buoys) alongside the main navigation channel in the estuary. These instruments measure the return signal or "backscatter" from suspended particles moving in the water column, which are assumed to be traveling at the same velocity as the water. They are mounted in a downward-looking configuration just below the water surface to resolve as much as possible of the vertical profile of the water currents. The instrument design allows for the selection of the width of vertical "slices" in the water column over which the water velocity is measured. In the deployments here, a 0.5-m slice or "cell" size is employed.
- One High-Frequency Surface Wave RADAR system, operated in tandem with two Rutgers University RADAR systems to provide broad-area measurement of surface currents and waves. These RADAR systems were developed by CODAR Ocean Sensors. Each system consists of a transmit antenna, a receive antenna, and hardware housed within a climate-controlled enclosure. The system employs the Bragg scattering of radio waves off of ocean waves in order to provide an estimate of wave magnitude and (using two or more arrays) the wave direction. The bandwidth of the radar sets the spatial range resolution of the system.
- Commuter ferry-based salinity and temperature sensors.
- Six anemometers providing detailed and continuous observations of local meteorological conditions.

Figure 12.3 provides a mosaic of photos of the various types of measurement systems employed in NYHOPS. The figure also provides a sense of the complex, challenging urban ocean environment in which these systems are deployed.

The shore-based and moored data are sampled every 10 minutes and are transmitted on an hourly basis via a radio link. The ferry data are sampled and transmitted every 30 seconds via a cellular connection. The land, moored, and ferry-based sensors transmit data via secure wireless networks. Data QA/QC procedures are in place to ensure the reliability of the observations. The data are also transferred to the NOAA National Weather Service's National Data Buoy Center for distribution via the National Data Buoy Center (NDBC) website (www.ndbc.noaa.gov/Maps/Northeast.shtml).

12.3.2   The Data Analysis-Modeling Subsystem

Computer forecast models have achieved great sophistication and accuracy over the past decade and are being used with success in providing multi-day forecasts for use in the assessment of future ocean and weather conditions by various user communities ranging from the marine transportation community

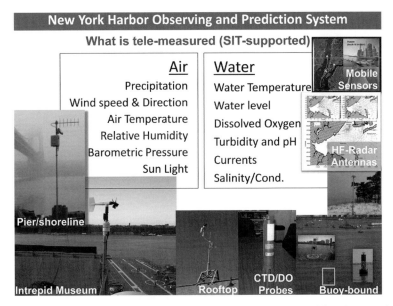

Figure 12.3 A photo mosaic of the measurement system types and locations employed in the NYHOPS ocean and weather measurement system (courtesy Stevens Institute of Technology). (A black-and-white version of this figure appears in some formats. For the color version, please refer to the plate section.)

to commercial and recreational fishermen. Computer models are also being used to great effect for the purpose of filling observation gaps in the space and time information domain, using data assimilation combined with "nowcasts" to provide estimates of ocean state variables at locations and times where no direct measurements exist.

Computer models therefore facilitate both the delivery of real-time ocean and weather information and the forecast of this information in the future, normally out to 48 or even 72 hours. Models can also be useful in the reconstruction of past conditions and in the conduct of hypothetical scenario simulations, both of which can be valuable tools particularly in urban ocean regions where, for example, there is often a need for emergency response preparation and/or coastal ocean resource management.

Again drawing from the experience in the design and operation of NYHOPS, the hydrodynamic forecast model is based on the Princeton Ocean Model (Blumberg and Mellor, 1987) and its shallow water derivative model, ECOMSED (Blumberg et al., 1999). The NYHOPS forecast model is run to provide both a hindcast (-24 hours) and a forecast ($+48$ hours) of the hydrodynamic circulation and wave conditions in the coastal (<200 m deep), estuarine, and freshwater zones from coastal Maryland to Cape Cod, Massachusetts, with a very high-resolution sub-model in the New York Harbor region, including the area around

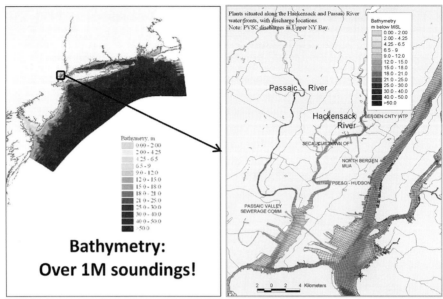

Bathymetry:
Over 1M soundings!

One high-resolution 3D model grid:150,680 water cells.

From 7.5 km to <50 m variable resolution. 7 coastal states.

Figure 12.4 NYHOPS computer model domain and high-resolution grid (courtesy Stevens Institute of Technology). (A black-and-white version of this figure appears in some formats. For the color version, please refer to the plate section.)

Manhattan. Ten layers are employed to define the vertical characteristics. The model domain and model grid are illustrated in Figure 12.4.

The hydrodynamic model is initiated at 0 hours local every day and completes a 24-hour hindcast cycle based on observed ocean and weather forcing followed by a 48-hour forecast cycle based on forecasted ocean and weather forcing. NYHOPS provides forecasts for water level, 3D circulation fields (currents, water temperature, salinity, water density), and wave characteristics (height and period). The hindcasts and forecasts are delivered four times per day. The NYHOPS model was extensively validated against water level, wave, water temperature, water salinity, and ocean current observations for a 2-year validation period.

12.3.3   The Data Management and Communications Subsystem

A reliable and sustained ocean and weather observing system cannot exist without the capacity to access, verify (quality assurance/quality control), and combine data across multiple information types and sources. When successful, users from across the various communities of scientists, the maritime industry, government and private decision-makers, regulatory officials, resource managers, and the general public are all able to search for and retrieve the data that they need, with confidence in the accuracy and (importantly) the limitations of the data. In practice, this requirement leads to the development and implementation of recommended or required standards and protocols.

Ideally, and as envisioned in the aforementioned US IOOS program, a fully functioning data management and communications system should provide the ability to collect and distribute both individual (tailored) data sets and metadata. The system should enable a uniform approach to quality assurance/quality control, although this will require a consistent approach to QA/QC across multiple observing systems, something that has not yet been achieved. The ideal system should also facilitate the archiving and reuse of data, including vastly different data types describing the physical, biological, and chemical characteristics of the coastal ocean. As the fields of data analytics and data visualization continue to expand both in terms of the ability to treat very large datasets and in terms of the sophistication with which complex data can be displayed and analyzed, researchers from across disciplines will perceive linkages among ocean state variables that had previously been hidden.

The NOAA CO-OPS (Center for Operational Oceanographic Products and Services), NDBC, and the US IOOS Office have been collaborating for several years on a data integration project that aims to increase interoperability between data providers and the user community. Among many outcomes was the establishment of the Quality Assurance/Quality Control of Real Time Oceanographic Data program (see: https://ioos.noaa.gov/project/qartod/). There are similar ongoing activities taking place around the world to ensure that ocean and weather data are collected, quality is assured, and the data are converted to data products useful to a wide range of users. The Met Office of the United Kingdom, for example, has developed a very wide range of data products (see: www.metoffice.gov.uk/services).

12.3.4   A System of Systems

In a very real sense, an advanced coastal ocean observing system is in fact a system of systems. For example, NYHOPS is composed of various modules, each working in tandem with the other and exchanging information to produce their individual NYHOPS products. The NYHOPS system of systems is comprised of the following modules:

- The oceanographic and meteorological data repository
- The NYHOPS forecast module
- The NYHOPS website and user-tailored products (Assimilated Nowcast, Storm Surge Warning System, Google Earth ® NYHOPS viewer, Coastal Inundation Mapper, etc., all viewable at http://hudson.dl.stevens-tech.edu/maritimeforecast)

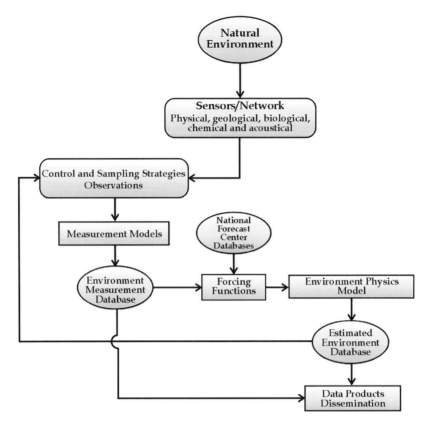

Figure 12.5 A flowchart of the NYHOPS system of systems.

Figure 12.5 illustrates a flowchart of the original NYHOPS system-of-systems design, indicating the various processes and outputs discussed herein.

In our view, only a system-of-systems approach with three major subsystems as described here can provide information related to the state of the coastal ocean and atmosphere (past, present, and future) to a known degree of accuracy. Advances in data acquisition and management, and in numerical modeling have allowed for the real-time use of data in improving model skill (via data assimilation techniques) and the use of the forecast model to significantly improve and expand our understanding of present conditions by producing data assimilative "nowcasts". The delivery of actionable present conditions, reliable forecasted conditions, and archived historical conditions all add up to a system that represents an indispensable tool for the many urban ocean user communities.

12.4   Unmanned and Autonomous Sampling Systems

Up to this point, we have described the design and operation of sophisticated coastal ocean observing and forecast systems that rely on the sustained, continuous measurement of a number of ocean state variables. By definition, these discrete measurements are fixed at known locations, described by the position of the mooring system, the shore-based mounting system, the position of remote sensing systems such as HF RADAR systems, and even the route of vessel-mounted sensors. In all cases, we know the precise location of the measurement, and we use this knowledge to enable data assimilation into forecast models and to generate continuous time histories of data.

Recent additions to the sensor platforms available to the coastal ocean observing community include unmanned vehicles capable of carrying a number of ocean sensors for extended periods of time. Gliders are submersible vehicles that use buoyancy to traverse an ocean region while "bouncing" up and down in the water column, producing a sawtooth-type pattern as they move forward. Gliders typically carry instrumentation to measure ocean state variables such as temperature, conductivity (salinity), and depth. Some have been equipped to also measure ocean currents, chlorophyll fluorescence, optical backscatter (turbidity), and acoustic backscatter. Ballast (either water or oil) is either brought into the vehicle or expelled in order to produce a downward or upward motion, respectively. The trajectory of the glider in the horizontal plane is controlled using either a rudder or a roll controller. Communication with the glider for the purpose of changing the route is accomplished when the glider is at the surface, via satellite or radio link. These surface intervals also allow the glider to transmit the acquired data and its GPS-based location. These simple but robust platforms have been in increased use around the world because of their ability to navigate large distances virtually unattended.

Another platform that has seen increasing use worldwide is the Wave Glider, a submerged vehicle that is tethered to a surface "mother ship" whose purpose is to generate power from surface wave action in order to provide the energy source for the propulsion, controls, and sensors on the submerged vehicle. The Wave Glider surface platform is also equipped with solar panels to provide propulsion in low-wave conditions and extra propulsion to navigate during periods of high opposing surface currents. The solar panels also recharge the batteries that power the sensors located on the submerged glider. The Wave Glider system can theoretically remain at sea for approximately 1 year, depending on the ocean conditions, and assuming of course that the control system and sensors remain functional, and that the system does not experience biofouling or other such complications. Satellite and/or radio communications allow for navigation control, location finding, and data transfer. The significant powering capabilities of the platform allows for a large payload and/or towing capacity (up to 500 kg), thereby allowing for the inclusion of a significant instrumentation payload (www.liquid-robotics.com/platform/how-it-works).

To-date, these unmanned systems have found great use in the provision of continuous data (e.g., temperature and salinity) while transiting very large distances in the coastal ocean, including along and across large stretches of the continental shelf. Perhaps most significantly, they have been extraordinarily useful in providing data during conditions that preclude the presence of human-operated systems, such as during the passing of a hurricane or during an oil spill. Clearly, these systems are now a permanent feature of the coastal ocean observer's toolbox. But they are not yet truly autonomous because they rely on the near-constant intervention of human controllers.

We here predict that an even more impactful future application of these systems and future systems like them, will be their use as a component of an integrated observing and modeling system such as the NYHOPS system described herein, but with true autonomy. As we continue to develop an understanding of the precursors to significant coastal ocean events such as Harmful Algal Blooms, we can imagine a future where the stationary sensors or even the forecast model provide an alert to an autonomous mobile system to navigate toward a location of interest in order to provide an immediate assessment of the conditions. This type of a response might also be driven by the stationary sensors when they detect a stark anomaly such as very high turbidity that might infer that a pollution event such as an oil spill has occurred; again directing the autonomous mobile system to respond.

12.5   Satellite-Based Observing Systems

Although we might view satellite-based observations of the ocean as a relatively recent advancement, these now-indispensable tools actually date back to the late 1970s! Seasat was launched by the National Aeronautics and Space Administration (NASA) on June 27, 1978, and operated until October 10, 1978. During that brief, 106-day period, Seasat carried the following instruments:

- Radar altimeter designed to measure the distance from the satellite to the ocean surface
- Microwave scatterometer to measure near-surface wind speed and direction
- Microwave radiometer to measure the ocean surface temperature
- Visible and infrared radiometer to distinguish between land and water areas on Earth, and to monitor cloud cover
- Synthetic Aperture Radar (SAR) to measure the ocean wave characteristics. The SAR also provided some of the earliest synoptic measurements of the sea ice cover at the poles.

More than a decade later, the next highly successful ocean observing satellite was launched on August 10, 1992, when TOPEX/Poseidon was launched. This was a joint mission between NASA and the French space agency. For more than 10 years, until October, 2005, the satellite provided the following measurements:

- Altimeter-based measurements of the position of the ocean sea surface, including a synoptic view of tidal variations around the globe
- Ocean surface temperature
- Using these synoptic, global measurements, maps of large-scale ocean currents and the description of other ocean dynamics such as Rossby waves, El Niño, and La Niña

Satellite-based sensing of the ocean, including the coastal ocean, is now an essential tool for ocean scientists, engineers, natural resource managers, and the coastal ocean user community ranging from ship navigators to fishermen and recreational boaters. Satellites now provide synoptic measurements of global sea surface temperatures at sufficiently high spatial resolution to monitor coral bleaching events and to identify in near real-time the onset of upwelling events, the details of critical large-scale currents such as the Gulf Stream, and the trajectory of river plumes along the coast. Satellite-based measurements of sea surface color, after years of data analysis and comparison with in-situ measurements at the ocean surface, now provide information regarding the presence of Harmful Algal Blooms and assist scientists in the differentiation of water types, e.g., river water, nearshore coastal ocean water, and deep ocean water. Satellite-based altimeters with centimeter-level accuracy are being employed, along with ground truth data and advanced analysis techniques, to track changes in global sea level. These highly accurate altimeters can also be used in the measurement of ocean surface waves across very large swaths of the ocean, as well as the mapping of large-scale topographic features beneath the ocean (via gravity anomalies) and large-scale ocean currents. Satellite images, now such an everyday commodity for the general public, help weather forecasters, navigators, and emergency management personnel to monitor the evolution and track of major weather systems including tropical and extratropical storms. For further information, the reader is referred to https://oceanservice.noaa.gov/facts/satellites-ocean.html and www.star.nesdis.noaa.gov/sod/lsa/.

12.6   Future Challenges

The last 15 years have provided numerous lessons in the value of the integration of ocean and weather observation systems at regional and global scales. We have seen partnerships develop among governments and universities worldwide, with best practices and lessons learned being shared broadly. New technologies and processes continue to be developed for data acquisition, QA/QC, and dissemination, and for high-resolution and data assimilative forecast modeling systems. We have witnessed dramatic improvements in these systems, including in the prediction of coastal hazards such as tropical and extratropical storms and storm surges. As new sensor technologies and new data analysis and visualization tools continue to be developed at a rapid pace, we can expect significant improvements in the associated products delivered to the user communities.

The improvement and expansion of coastal ocean and weather observation and prediction systems around the world will require continued efforts to ensure integration and interoperability of these systems across organizations that include private, public, and academic entities. These partnerships help to ensure that data products and services are responsive to the needs of the various user communities, from scientists and natural resource managers, to the maritime community and recreational users of the coastal ocean. This in turn drives support for the continued investment in public and private resources for the operation and maintenance of the systems. This process will require trust-building among all parties, including the user communities. And it will require that the partners address the sometimes complex issues related to data owner-ship and data product liability.

Many challenges remain, including the following.

- Integration of biological data
  A primary aim of urban coastal ocean observation systems is the improved understanding of the state of the coastal ocean ecosystem and its inhabitants. Ultimately, this understanding should be expanded to include the causative factors that link human activities to the health of the ecosystem. Sensor and data analysis systems are needed to provide accurate assessments of marine species abundance, life stage, behaviors, etc. The integration of these data with physical and chemical data would provide powerful tools for use by the broad scientific community, natural resource managers, policy makers, and educators.
- The explosive growth in data
  The last 20 years have seen remarkable advances in sensor technologies, including both in-situ sensors such as ocean current profilers, remote sensing technologies such as RADAR and satellite-based sensors, and mobile, semi-autonomous sensor systems mounted on surface and underwater vehicles. As a result, there has been an associated remarkable increase in both the volume and type of data available to the scientific

and user communities. Significant effort will be required to develop technologies and processes to handle the enormous volume of data. In addition, collaborative efforts will be required to better understand the data associated with new and emerging sensors, and the ways in which the data from these sensors can be effectively integrated into existing, well-understood data streams.

- Testbeds

  There has been a great deal of work toward establishing standards for new sensing technologies, sampling protocols, data analysis, and data QA/QC. There has not however been significant progress in the development of testbeds and testing protocols and procedures to examine the operational performance of individual systems and integrated systems in a realistic environment. In order to be effective, these testbeds should be fully inclusive and transparent, and they should encourage the public–private–academic collaboration necessary for long-term sustained success.

- Education and training

  There is an urgent and ongoing need to ensure that we as a community commit to the education and training of the next generation of ocean and weather observing and modeling professionals. Operational systems ultimately are as effective, robust, and resilient as the people who design, deploy, maintain, and operate them. These are highly specialized skills gained only by doing. And we must recognize that many of the requisite skills, including sensor and vehicle design, data analysis, computer programming, and systems engineering, are highly valued in other fields. Certainly, the integration of real-time coastal ocean and weather observing systems into K-12 STEM education provides opportunities to inspire young people to pursue careers associated with ocean observing.

13

Climate Change

13.1 Introduction

The urban coastal regions of the globe are particularly vulnerable to a changing climate in that they are immobile. Animals and people can move from their locations if they become uncomfortable. Urban infrastructure such as bridges, subway systems, buildings, and roads, and deep-rooted residents cannot. These strengths of place can, however, become liabilities if the local ecosystems that they are based on are unable to adapt to the climate-induced changes. Climate change poses serious threats to urban infrastructure, quality of life, and entire urban systems. Not only poor countries, but also rich ones will increasingly be affected by anomalous climate events and trends.

Urban areas are both greatly affected by the impacts of climate change and major contributors to the emission of greenhouse gases that are causing the changes in the first place. The world is at a crossroads: inaction will reduce citizen welfare, increase costs and insecurity, and eventually risk urban catastrophe as resources are depleted and climate damages mount. Resource- and carbon-efficient growth is the only sustainable long-term option. The choices made in cities today on transport, infrastructure, buildings, and industry, as they grow rapidly over the coming decades, will determine the technology, institutions, and behaviors they lock in and govern whether mankind can both manage climate change and capture the benefits of resource-efficient growth.

Urban coasts face special climatic conditions that must be accounted for when preparing for any adaptation plans due to a warming climate. These include:

> Urban heat island – Cities already tend to be hotter than surrounding suburban and rural areas

Air pollution – The urban boundary layer's capping inversion that traps pollutants from residential, commercial, industrial, and transportation activities

Flooding – Hazards from heavy downpours, and coast and riverine flooding already plague the urban coasts

This chapter examines physical phenomena observed in the Earth's weather and climate with a focus on the urban coasts, providing sufficient scientific and technical background to enable readers to critically examine arguments being discussed by policy makers and the public at large. The focus will be on the physical impacts that affect cities the most: temperature, precipitation, sea level rise, and extreme events.

13.2   Signs of Change

The world has warmed over the last 150 years, especially over the last six decades, and that warming has triggered many other changes to Earth's climate. Evidence for a changing climate is seen everywhere, from the top of the atmosphere to the depths of the oceans. The documented changes in surface, atmospheric, and oceanic temperatures come from melting glaciers, disappearing snow cover, shrinking sea ice, and rising sea level (Figure 13.1). In 2016, Earth reached its highest temperature on record, beating records set in 2015 and 2014. Global annual average temperature has increased by more than 0.7°C (1.2°F) for the period 1986–2016 relative to 1901–1960. The increase is not smooth over time because the warming trend is superimposed on natural variability of the Earth's climate.

Now 0.7°C at first blush does not seem like much given the large changes that occur from summer to winter, but the global distribution is not uniform (Figure 13.2). Warming occurred mostly in the high latitudes of the Northern Hemisphere with changes of 2°C. In 2003, a heat wave struck Europe in which the average temperature in Europe was 3.5°C above average, leading to more than 70,000 deaths (Robine et al., 2008) from extreme heat related stresses. Most ocean areas around Earth are warming too. All US coastal waters have warmed by more than 0.4°C (0.7°F). The ocean is expected to warm more slowly given its larger heat capacity, leading to land–ocean differences in warming (as seen in Figure 13.2). As a result, the climate for land areas often responds more rapidly than the ocean areas, even though the forces driving a change in climate occur equally over land and the oceans.

Observations from tide gauges indicate that global mean sea level has risen about 20–23 cm (8–9 inches) since 1880, with a rise rate of approximately

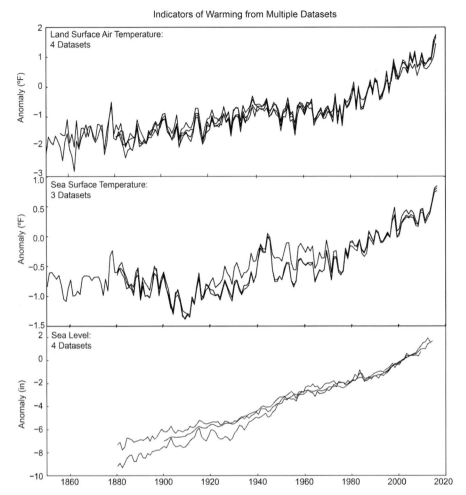

Figure 13.1 Global observations of land surface air temperature (1850–2016 relative to 1976–2005, increase); sea surface temperature (1850–2016 relative to 1976–2005, increase) and sea level (1880–2014 relative to 1996–2005, increase) (USGCRP, 2017).

12–15 mm/decade (0.5–0.6 inches/decade) from 1901 to 1990 (Figure 13.1). However, since the early 1990s, both tide gauges and satellite altimeters have recorded a faster rate of sea level rise resulting in about 8 cm (about 3 in) of the global rise since the early 1990s. On a more local urban ocean scale, observations in New York Harbor (Figure 13.3) show that the annual maximum water level (tides plus storm surge) has increased markedly. Three of the nine highest recorded water levels have occurred since 2010 and eight of the largest twenty have occurred since 1990. The increases are due to (a) relative mean sea level rise

Surface Temperature Change

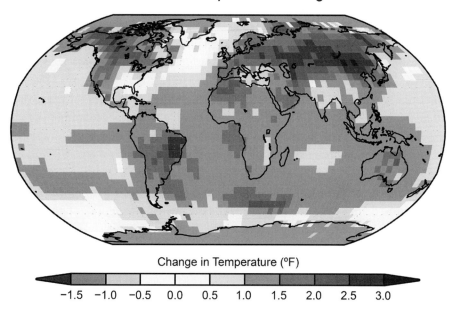

Figure 13.2 Surface temperature change (in °F) for the period 1986–2015 relative to 1901–1960 (USGCRP, 2017). (A black-and-white version of this figure appears in some formats. For the color version, please refer to the plate section.)

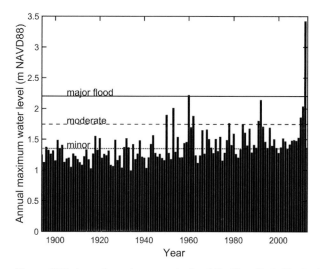

Figure 13.3 Annual maximum water level for New York City (at The Battery, or nearby stations for 4 missing years), compared with present-day minor, moderate, and major flood thresholds from the National Weather Service (redrawn by Philip Orton, 2017, with data from Talke et al. [2014], as corrected to add the annual mean sea level under each year's maximum storm tide).

and (b) possibly storm changes or harbor changes (e.g., dredging, landfill). At present, virtually every year has at least minor flooding.

Traditionally, cities were located near rivers and oceans for transportation and connectivity purposes. This natural geographic advantage is now increasing vulnerability of cities as sea levels rise and significant wind events increase in severity and frequency. In Europe, 70% of the largest cities have areas that are particularly vulnerable to rising sea levels, and most of these cities are less than 10 m above sea level. Port cities in emerging or relatively new economies such as Kolkata, Shanghai, and Guangzhou are as vulnerable as such cities in developed countries that have been highly developed for many decades, e.g., Rotterdam, Tokyo, or New York City. Already global sea level rise and its regional variability is contributing to significant increases in annual tidal-flood frequencies, which are measured by NOAA tide gauges and associated with minor infrastructure impacts to-date; along some portions of the US coast, frequency of the impacts from such events appears to be accelerating.

13.3   Drivers of Climate Change

The Earth's climate is a complex physical system. Nevertheless, we can still understand much about the climate even without an advanced degree in physics or atmospheric science. Most climate scientists agree the main cause of the current global warming trend is human expansion of the "greenhouse effect". Life on Earth depends on energy coming from the sun. About half the light reaching Earth's atmosphere passes through the air and clouds to the surface, where it is absorbed and then radiated upward in the form of infrared heat. About 90% of this heat is then absorbed by the greenhouse gases and radiated back toward the surface, which is warmed to a life-supporting average of 59°F (15°C). When the amount of greenhouse gases increases, the amount of heat absorbed by the gases increases preventing more and more heat from radiating from Earth toward space.

There are certain gases in the atmosphere that block heat from escaping. These so-called greenhouse gases contain molecules of mainly water vapor (H_2O), carbon dioxide (CO_2), methane (CH_4), and nitrous oxide (N_2O), all acting as effective global insulators. To explain how, let us use CO_2 as an example. When its atoms are bonded tightly together as they are for a CO_2 molecule the molecule will absorb infrared radiation and this will in turn cause the molecule to start to vibrate. Due to this vibration, the molecule will eventually emit the radiation again, and this radiation is likely to be absorbed by another molecule, and so on. This exchange of heat will go on and keep the atmosphere filled with these vibrating molecules as an insulating surface around the globe.

The reason why some gases are worse than others is because of the bond in-between the molecules. It needs to be tight enough for a vibration to start when they are affected by radiation. Gases formed by bonds between two oxide molecules and two nitrogen molecules, which are the two most common molecules in the atmosphere, are not greenhouse gases because of the loose bond in-between the molecules.

Long-lived gases that remain semi-permanently in the atmosphere and do not respond physically or chemically to changes in temperature are described as "forcing" climate change. Emissions of greenhouse gases come from energy production and consumption, agriculture, transport, and ecological processes. Gases, such as water vapor, that respond physically or chemically to changes in temperature are "feedbacks."

Gases that contribute to the greenhouse effect include:

- Carbon dioxide (CO_2). The burning of fossil fuels like coal and oil has increased the concentration of atmospheric carbon dioxide (CO_2) over the last century. The emission of gases from the burning of fossil fuels continues to be the major contributor to climate change accounting for 75% of the warming impact.
- Water vapor (H_2O). This is the most abundant greenhouse gas, but importantly, it acts as a feedback to the climate. Water vapor increases as the Earth's atmosphere warms, but so does the possibility of clouds and precipitation, making these some of the most important feedback mechanisms to the greenhouse effect.
- Methane (CH_4). A hydrocarbon gas produced both through natural sources and human activities, including the decomposition of wastes in landfills, agriculture, and especially rice cultivation, as well as ruminant digestion and manure management associated with domestic livestock. Methane is a far more active greenhouse gas than carbon dioxide but is much less abundant in the atmosphere. Methane accounts for about 14% of the warming impact.
- Nitrous oxide (N_2O). A powerful greenhouse gas produced by soil cultivation practices, especially the use of commercial and organic fertilizers, fossil fuel combustion, and biomass burning. This gas is present in much smaller quantities and its overall contribution to global warming is very small, about 8%.
- Chlorofluorocarbons (CFCs). Synthetic compounds entirely of industrial origin used in many applications, but now largely regulated in production and release to the atmosphere by international agreement.

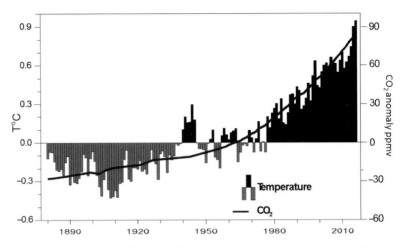

Figure 13.4 Time series of annual values of global mean temperature anomalies (bars) and carbon dioxide concentrations at Mauna Loa, both from NOAA. Data are relative to a baseline of the twentieth century values. From Trenberth and Fasullo (2013).

The most controllable of the greenhouse gases is carbon dioxide (CO_2). Humans have increased atmospheric CO_2 concentration by more than a third since the Industrial Revolution began. This massive increase is linked to continued growth in production and in international trade in goods and services. Increasing dependence on motor vehicles makes transportation one of the main sources of greenhouse gas emissions.

Agriculture is also a major source of greenhouse gases, accounting for about one-third of global emissions of carbon dioxide, methane, and nitrous oxide. This source of greenhouse gases is much more difficult to control as wealthy consumers in industrialized countries, and increasingly in emerging countries too, expect a wider choice in food products. Consumers are for example demanding year-round choices of fresh fruit and vegetables requiring more worldwide transportation of food.

An examination of annual values of global mean temperature anomalies and carbon dioxide concentrations as shown in Figure 13.4 over the past 135 years suggests a strong relationship between CO_2 and temperature. The relationship is not linear. In other words, while the CO_2 has steadily increased, the temperature tends to fluctuate in the short-term because there are important short-term natural events, such as El Niño and La Niña and variability in pollutant loadings from atmospheric aerosols and the occasional volcano eruption, that can either enhance or inhibit the effects of CO_2. Even with the short-term "noise", the longer-term increase in temperatures is in concert with the increasing CO_2.

Figure 13.4 is not definitive in establishing proof that CO_2 increases are related to the temperature increase but it is highly suggestive. Proof will come via the use of global climate models discussed next.

13.4   Forecasting Future Climates

Predicting how much greenhouse gases and aerosols will be emitted into the atmosphere each year from human activities is a challenge. Such predictions are known as emission scenarios. The emissions can be related to the factors that control it by

$$I = P \cdot A \cdot T, \tag{13.1}$$

where I is the total emissions of greenhouse gases impacting our climate, P is the population, A is the level of affluence, and T stands for the greenhouse gas intensity. Affluence is measured by the gross domestic product (GDP) per person, and T is related to the energy it takes to generate the goods and services and the efficiency with which the economy uses energy, a measure of technology. Equation 13.1 is commonly called the IPAT relation or the Kaya Identity. Although we have a good idea of the factors that control the total emissions, making estimates of the factors is difficult. Scientists then developed emission scenarios to account for the uncertainty in the estimates. The scenarios are called RCPs – Representative Concentration Pathways. The scenarios are then input to climate models, which calculate a future climate for each of them.

Global climate models use mathematical equations very similar to those derived in Chapters 4 and 5 to describe the behavior of factors of the Earth system that impact climate. These factors include dynamics of the atmosphere, oceans, land surface, living things, and ice, plus energy from the sun. Sophisticated climate models are increasingly able to include details such as clouds, rainfall, evaporation, and sea ice. Climate models are constructed with three types of essential building blocks: physical, chemical, and biological laws founded on theory and built with data. Climate models divide the globe into a three-dimensional grid of cells representing specific geographic locations and elevations. Each of the components (atmosphere, land surface, ocean, and sea ice) has equations calculated on the global grid for a set of climate variables such as temperature. In addition to model components computing how they are changing over time, the different parts exchange fluxes of heat, water, and momentum. They interact with one another as a coupled system. And of course they require the use of supercomputers and parallel computing. The models are used to simulate conditions over hundreds of years, so that we can predict how our planet's climate will likely change.

Unlike weather, which takes place in a matter of days, climate unfolds over decades; testing a climate model, therefore, is no simple task. Because it is impractical to wait 30 or 50 years to verify a model's accuracy, models are usually tested by their ability to recreate past climates. Models that produce results similar to what actually happened are considered to be relatively trust-worthy and give us confidence in using these models to project future climate change.

One criticism of climate predictions is that since we cannot predict the weather more than a week ahead, how can we state confidently what the climate will be in 2100? This criticism is built on a fatal flaw as it makes the mistake of equating weather predictions to climate predictions. These predictions are fundamentally different problems. A weather forecast for tomorrow, for example, needs to accurately know the state of the atmosphere today. Small errors in what we know of today's atmosphere will grow exponentially so that a forecast more than a few days into the future will be dominated by the errors in our knowledge of today's atmosphere.

Climate models do not require an accurate knowledge of the state of the atmosphere today. Instead they require a knowledge of the radiative forcing, the imbalance between incoming solar radiation and outgoing infrared radiation that causes the Earth's radiative balance to stray away from its normal state. This straying causes changes in global temperatures. A climate prediction is a prediction of the statistics of the system. For many complex systems, predicting the statistics is easier than predicting the specific state of the system.

13.5   A Warming Climate

The projected changes in globally averaged temperature for a range of emission scenarios that vary from assuming strong continued dependence on fossil fuels in energy and transportation systems over the twenty-first century (the high scenario RCP8.5) to assuming major emissions reduction (RCP2.6) are shown in Figure 13.5. These scenarios account for the uncertainty in future emissions from human activities (as analyzed with the 20+ models from around the world used in the most recent international assessment). The large emissions associated with the RCP8.5 scenario lead to temperature increases of 4°C (8°F) by 2100 while the low emissions associated with the RCP2.6 scenario leads to temperature increases of only 1°C (2°F) peaking by 2020 or so. To put these changes into perspective, it took a warming of about 4°C (8°F) to separate our modern world from the last ice age. We are warming the entire planet. It is also worth noting that the temperature trajectory over the next few decades has already been determined largely by greenhouse gas emissions that have

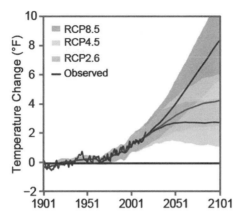

Figure 13.5 Multimodel simulated time series from 1900 to 2100 for the change in global annual mean surface temperature relative to 1901–1960 for a range of the Representative Concentration Pathways. The mean (solid lines) and associated uncertainties (shading, showing ±2 standard deviations [5%–95%] across the distribution of individual models based on the average over 1901–1960) are given for all the RCP scenarios.

already occurred. These existing emissions would commit the world to at least an additional 0.6°C (1.1°F) of warming over this century relative to the last few decades.

Although the global average temperature is projected to increase, it will not increase uniformly around the Earth. Figure 13.6 shows the distribution of warming predicted by the lowest emissions scenario, RCP 2.6. This scenario assumes immediate and rapid reductions in emissions and would result in about 1°C (2°F) of warming in this century. The highest scenario shown in Figure 13.7, RCP 8.5, roughly similar to a continuation of the current path of global emissions increases, is projected to lead to more than 4°C (8°F) warming by 2100, with a high-end possibility of more than 6°C (~11°F). As shown in Figures 13.6 and 13.7, continents are warming faster than the oceans because the ocean has a higher heat capacity than land. Warming in northern North America and Eurasia is projected to be more than 40% greater than the global average warming. And high latitudes will warm more than the tropics leading to greater ice melt and more sea level rise.

13.6   Precipitation Changes

It is well known that the global climate models used to project precipitation changes exhibit various degrees of fidelity in simulating the

Rapid Emissions Reductions (RCP 2.6)

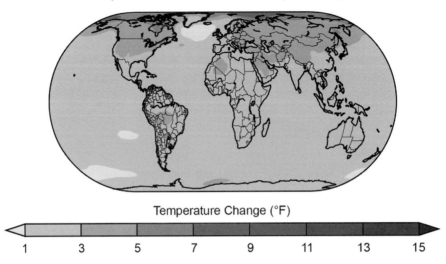

Temperature Change (°F)

Figure 13.6 The distribution of warming predicted by the lowest emissions scenario, RCP 2.6. (A black-and-white version of this figure appears in some formats. For the color version, please refer to the plate section.)

Continued Emissions Increase (RCP 8.5)

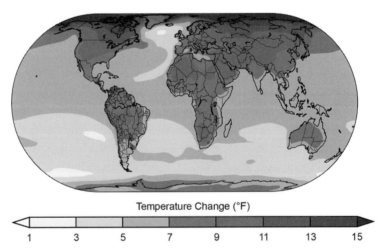

Temperature Change (°F)

Figure 13.7 The distribution of warming predicted by the highest emissions scenario, RCP 8.5. (A black-and-white version of this figure appears in some formats. For the color version, please refer to the plate section.)

observed climatology. We do know however that because of the increased greenhouse gases in the atmosphere, the trapping of infrared radiation will increase. This leads to an increase in evaporation of the oceans, and because precipitation must balance evaporation, precipitation will also increase.

We can say with confidence that the analysis shows that total global precipitation will increase by a few percent for every degree of global warming. The increase like the increase in temperature will not be distributed uniformly. There will be a large-scale shift of precipitation to higher latitudes causing a decrease in many parts of the tropics, and there will be less wintertime snow; instead there will be more wintertime rain. In the United States, the frequency and intensity of heavy downpours are projected to increase over the next 100 years.

13.7   Sea Level Rise

Sea level rise is one of the most certain impacts of climate charge and will pose a growing challenge to urban coast communities from increased inundation, more frequent flooding, and erosion of coastal landforms. The sea level rises because of a warming climate for two reasons. First, as ice on land melts, the melt water flows into the ocean increasing sea level. Note that floating ice does not contribute to sea level rise when it melts because the ice has already increased the water level by displacing its volume through its weight when it entered the water. This effect is known as Archimedes' Principle. Second, as the temperature of water increases it expands taking up a greater volume and increasing the sea level. The latest results from the New York City Panel of Climate Change (PCC) for New York Harbor and the Lower Hudson River (Figure 13.8) suggest that the middle range (25th–75th percentile) sea level rise projection is an increase of 0.15 m (0.5 ft) in the 2020s, 0.45 m (1.5 ft) in the 2050s, and 0.90 m (3 ft) by 2100. Sea level rise is projected under the RCP 8.5 scenario to accelerate as the century progresses and could reach as high as 1.8 m (6 ft) by 2100.

To provide some perspective on this sea level rise, consider the amount sea level would rise if all the people on Earth jumped into the ocean at once. The estimate is:

Population of Earth = 7.6 billion people = 7.6×10^9 people
Average weight of a person = 70 kg
Surface area of Earth (assuming 70% is water) = 3.6×10^{12} m^2
Density of water = 1,000 kg m^{-3}
Volume displaced = weight/density of water
Volume displaced = 70 kg person^{-1} × 76 × 10^8 people/1,000 kg m^{-3} =
$\qquad$ 5.320 × 10^8 m^3
Sea level rise = Volume displaced/Ocean surface area of Earth =
$\qquad$ 1.5 × 10^{-6} m or 0.0015 mm

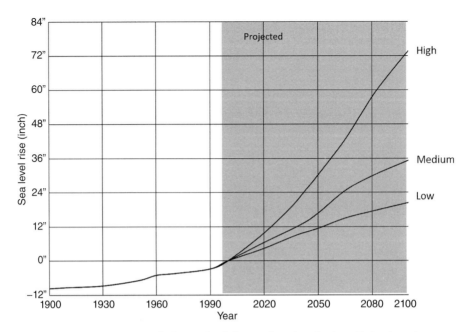

Figure 13.8 New York City sea level rise trends and projections. Projections shown are the low estimate (10th percentile), middle range (25th–75th percentiles), and the high estimate (90th percentile) (redrawn from Horton et al., 2015).

This tiny amount of 0.0015 mm suggests that humans do more damage out of the water by emissions of greenhouse gases than if all were to jump into the water.

13.8   Extreme Events

Extreme weather events are the most serious of the climate impacts because they increase people's vulnerability and make it even more difficult to adapt to a warming climate. While changes in the pattern of precipitation will occur, it is likely that more rainfall will come in heavy downpours. This will increase the risk of flooding and the loss of freshwater for humans and ecosystems because the water runs off quickly rather than soak into the soil. Hurricanes and their impacts will change too. The intensity, frequency, and duration of North Atlantic hurricanes, as well as the frequency of the strongest (Category 4 and 5) hurricanes, have all increased since the early 1980s. The relative contributions of human and natural causes to these increases are still uncertain. However, we can say with certainty that hurricanes will get stronger since their source of energy, the ocean surface temperature, will be increasing. The impacts of hurricanes will be more severe because a rising sea level will lead to higher storm surges and flooding as well as increased rates of coastal erosion.

Figure 13.9 A benefit to sea level rise? (Cartoon by Nick D. Kim, scienceandink.com, used with permission.)

13.9   National Security Concerns

Climate change will affect countries in different ways and to different extents. Some areas will become uninhabitable and that will cause people to move to areas that are, for example, less dry, less affected by large natural disasters (such as frequent stronger hurricanes), or have less flooding, which are all consequences of global warming. Some countries will probably start to argue that the climate refugees will become a national security problem. With climate change comes the effects of resource scarcity, extreme weather, food scarcity, water insecurity, and increased flooding from a rising sea level. These effects can lead to instability within a country or between countries.

To ensure national security, the governments will have to come up with policies and plans to provide resiliency to the effects of climate change. With resource scarcity, prices will sky rocket, tensions with countries vying for the resources will increase, and potential wars could break out. No one knows how instability will affect the world, but everyone knows that it will bring changes to the global community as well as within individual countries. Extreme weather brings more devastating natural disasters that require money, time, and labor to fix. More hurricanes during the season, or more nor'easters, will take a toll on economies, people, and their homes. Overall, the effects of climate change lead to harder times and less supplies or resources to go around. The unrest and instability in response to the effects of climate change threaten our national security and require the government to prepare to respond to threats and other events.

13.10   Are there any Possible Benefits?

There are a number of certain impacts of climate change. The climate will get warmer with more extreme heat events, precipitation patterns will change, sea level will rise, and the waters of the coastal and open ocean will get more acidic making it more difficult for ocean organisms to extract carbonate from the water to build their shells and skeletons. The impacts of these changes on human society will not be distributed evenly either. The wealthier countries will have an easier time adjusting to the changes than the poor countries will. But as the warming continues even the richest countries will be severely challenged. There may be some people and biota who will benefit from a warming climate (see the cartoon of Figure 13.9 as to the benefits of sea level rise), particularly if the warming is small, but overall there are serious negative impacts that should compel our attention.

14

Cities and Water: Building Resilience

14.1 Introduction

Here we seek to understand the vulnerability of coastal urban areas to natural hazards and the ways in which we can reduce this vulnerability through informed urban design. Vulnerability is often miscalculated, in part because of uncertainties in our knowledge of where and when extreme events might occur and in part because of a lack of understanding of the consequences of a particular event, remembering that an event of the type and magnitude being considered may not have ever occurred at the location. As has already been discussed, risk can be expressed as the multiplication of the probability of a particular hazard times the consequence of the hazard should it occur. The reduction of risk in a given urban area can therefore be accomplished by reducing the probability of occurrence of a known hazard and/or by reducing the consequence of that hazard. This leads naturally to a discussion of mitigation and adaptation. It also leads us to the concept of *resilience*, whereby we recognize the limits of our ability to reduce the likelihood of certain hazards and so focus our attention on learning and adapting in a way that can facilitate the rapid restoration of community functioning in the aftermath of a disaster.

Resilience describes the attributes of a system (e.g., a transportation system) that enables it to withstand, respond, and adapt to a vast range of disruptive events by preserving and even enhancing critical functionality. In other words, a resilient system ensures a high level of performance even under unexpected conditions. Challenges to resilience include 'external' threats from a range of hazards, and 'internal' threats from organizational deficiencies. As a result of lessons learned during and after extreme events around the world, many of them mentioned in this book, we do understand that there exist engineered solutions to the improved resilience of socio-technical systems. Increasingly, however, we also understand that the development of these solutions will require a transdisciplinary approach that includes not only engineering, planning and design, but also the natural, physical, and social sciences; economics; and policy. Many of these solutions will require assessment and predictive capabilities that do not presently exist, including the identification, collection, and analysis of relevant data (Bruno et al., 2015).

Our interest here in the improvement of coastal community resilience leads to an interest in exploring aspects of urban design that focus on the sustainment of community functions and critical urban assets while also creating social benefits. We need to develop an understanding of the increased risk of disaster associated with urbanization and development, coupled with the effects of climate change. Many urban areas already have substantial vulnerabilities, including substandard infrastructure and building environments, and socio-economic inequalities. It is important to address disaster risk from the standpoint of hazard exposure and vulnerability, with the aim of increasing the capacity for preparedness. Many if not most of the communities most at risk are also the least equipped to address the risks.

The need to better understand risk and resiliency in and among complex systems has been a strong feature of recent discussions among urban planners, policy makers, and emergency preparedness professionals. An emergent view is that resiliency is often about people, community, and education. There is also an emerging and broad consensus that urban areas and cities present uniquely challenging socio-economic and socio-technical issues with regards to their resiliency. Cities located in emerging economies do not presently possess the capacity to address many of the issues discussed herein, placing them at a disadvantage. However, cities located in the developed world, while possessing the capacity to act, are at the same time disadvantaged by virtue of their significant investments in legacy infrastructure that will require very costly and complex retro-fitting to provide enhanced resilience to the range of known threats.

In this chapter we discuss the emerging tools and supporting activities that can support design for resilience. The concept of aligning politics, finance, and design becomes central to this discussion. The roles of the social sciences and community engagement are introduced since risks and opportunities are too often not equally shared across the socio-economic landscape.

14.2   The Context of Change

It is important to recall our prior discussions about the fact that the threats to coastal urban populations have increased considerably over recent decades, as the probability of occurrence of natural hazards has increased (e.g., via the effects of climate change) and the consequence of these events has also increased (e.g., via population migration to the coast). Figure 14.1 puts this element of change into further historical perspective. The map of trade routes from 1912 identifies the major port cities of the time. Note how few of today's coastal megacities appear on the map, illustrating how rapidly population growth, migration patterns, and perhaps most significantly economic globalization over the last hundred years has profoundly changed the landscape of coastal population vulnerability and the need for a global conversation about resilience. Just as important, this map reminds us that efforts to ensure the resilience of socio-economic and socio-technical systems worldwide must be ongoing and must reflect the latest in our understanding of both the hazards and the approaches to mitigate these hazards and preserving/enhancing functionality under all possible conditions.

14.3   Designing for Resilience

The term resilience has been in use for many years to describe the behavior of a vast range of phenomena, including ecological systems (e.g., Holling, 1973), physical systems (e.g., structural dynamics), complex infrastructure systems (e.g., the supply chain), and communities. Certainly, there is still some debate about the most appropriate definition of the term "resilience". Norris et al. (2008) provide a useful summary of the applications of the term over the last 40 years across a variety of applications, and at scales ranging from a single individual to ecological systems. They conclude that in the context of human communities, organizations, and societies, resilience can best be defined as "a process linking a set of adaptive capacities to a positive trajectory of functioning and adaptation after a disturbance." The definition of resilience adopted by the Rockefeller Foundation's 100 Resilient Cities initiative (to be discussed later in this chapter) is "resilience is the capacity of individuals, communities, institutions, businesses, and systems within a city to survive, adapt, and grow no matter what kinds of chronic stresses and acute shocks they experience." (www.100 resilientcities.org). It is interesting to note that this definition includes the term "grow", thereby adding an element of improvement or wellness to the definition of resilience. In our view, these definitions can be summarized as the ability of a community and its supporting socio-economic and socio-technical systems to respond, learn, and adapt in order to preserve and even enhance functionality under both expected and unexpected conditions.

In view of the fact that we are here considering both the built environment and the natural environment, any possible success will require that we bridge the

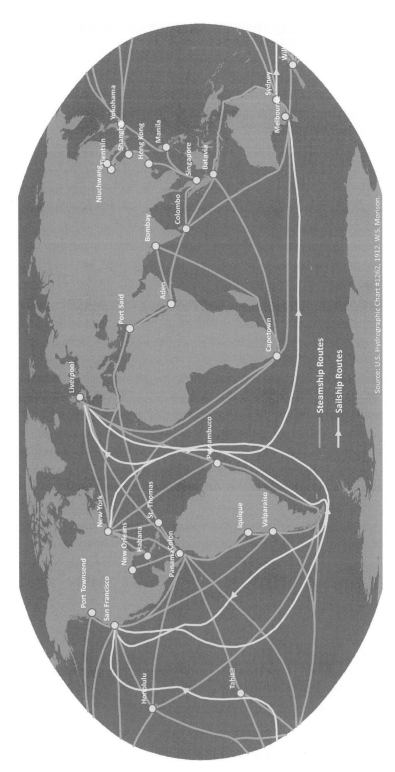

Figure 14.1 World Trade Routes in 1912 (Morison, 1912).

Niuchwang Tientsin Shangl Yokohama
Hong Kong
Manila
Singapore Batavia
Colombo
Bombay
Aden
Port Said
Liverpool
Pernambuco
New York
St. Thomas
New Orleans Habana
Panama Colon
Iquique
Valparaiso
Port Townsend
San Francisco
Honolulu
Tahiti
Capetown
Sydney
Melbourne
Wil

Steamship Routes
Sailship Routes

Source: U.S. Hydrographic Chart #1262, 1912. W.S. Morison.

gaps among the fields of engineering, planning, design, the natural, physical and social sciences, economics, and policy. Our experience in addressing the resiliency challenges faced by coastal communities indicates that the primary challenges to resiliency include: (1) uncertainty; (2) complexity; (3) the lack of capacity; and (4) globalization. Let's here examine each of these challenges.

Uncertainty refers to the evolving threats posed by natural hazards as described herein, as well as the evolution of geopolitical conditions and technology, and the response of communities to these changes. The uncertainty associated with both the threats and the range of responses must be better understood and better articulated to policy makers, decision-makers, and the general public. As has already been mentioned, progress toward better preparation for an extreme event is often hampered by the fact that the difficult economic and policy decisions are often only possible after an extreme event has already taken place – witness the significant activities underway in Puerto Rico following Hurricane Maria in September, 2017. A new approach must be adopted, informed by the best available science for predicting and articulating in easily understood terms the likelihood of potential coastal hazards, **along with** the uncertainty associated with such predictions. The ability to clearly understand and articulate uncertainty will lead to the essential attribute of adaptability that is so important to community resilience.

Complexity refers to the complexity of emerging technologies and the interactions and interdependencies that exist across different critical infrastructure systems, socio-technical systems, and socio-economic systems. During Hurricane Sandy in New York City, for example, the loss of power resulted in the inability to deliver fuel, which in turn hampered the delivery of food, water, and other critical supplies, as well as the operation of generators at critical medical centers and other urgently needed facilities. The direct delivery of fuel by barge and trucks was severely limited by local and national regulations that required many levels of approval and resulted in delays of multiple days. A system-of-systems approach is needed, with the ability to simulate and predict the various possible "failure cascades" that often occur during extreme events. These tools – and the use of rapidly improving visualization technologies – can dramatically improve the ability of communities and their leaders to prepare, respond, and adapt to virtually any imaginable event. It is entirely possible that this adaptation may in fact prompt planners and designers to seek to lessen the complexity of certain components of critical infrastructure. For example, life-sustaining equipment in hospitals could be designed to function entirely on battery power or even have the option of operating on human power!

Capacity refers to both the technical skills and the management systems to develop and implement resilience strategies tailored to a specific location and a specific set of possible threats. The lack of capacity is unfortunately often most severe in areas that are most in need and yet do not have the necessary human and financial resources. The greatest needs are more effective technology transfer and knowledge transfer, as well as innovative education and training systems. An important consideration in all of this is the fact that we are often speaking about preparing for and adapting to events that have not yet occurred, making the development of training systems all the more challenging. It is our strong opinion that the rapidly developing fields of data analytics and visualization can be a strong contributor here, perhaps including virtual reality and immersive environments that can provide the opportunity for planners, designers, and decision-makers to observe firsthand the consequences of a particular event in their city. Providing the opportunity to observe the event under multiple scenarios, for example with and without proposed mitigation and/or adaptation strategies, would be a powerful additional capacity to coastal communities worldwide.

Globalization refers to the fact that our socio-economic and socio-technical systems are global in their construct and in their functioning. In a very real sense, risk is increasingly shared across national boundaries and so solution paths must be developed and shared across national boundaries. These solution paths can and should include metrics for resilience, as well as standards and best practices to guide planners, designers, and decision-makers. Information-sharing must occur across public, private, and academic organizations, in a fashion that can protect privacy and proprietary information. An example of how effective information-sharing and communication on a global scale enhanced the resilience of an urban ocean community occurred in New York City during Hurricane Sandy. When it became clear that the Port of New York and New Jersey would be closed by the hurricane for an extended period of time, shippers, carriers (vessel operators), and their customers worked together to divert time-critical shipments away from the New York region and to the ports at Norfolk, Virginia, and Halifax, Canada. From those locations, cargo was loaded on to trains for delivery to the New York metropolitan region.

Clearly, these challenges can be used to describe a broad set of initiatives that could be undertaken within the planning and design communities, as well as within the research and education and training communities. One potential obstacle to this effort is the fact that the outcome – resilience – is not easily quantified, standardized, and measured. As mentioned earlier, different research communities have employed different definitions of resilience. Perhaps as a result, most of the academic work in resilience to-date has been qualitative. No common language or underlying theory and quantitative rigor has been developed. Part of what is required is a deeper understanding of the concept of community and the characteristics of community resilience. Indicators must be developed to define metrics of resilience and identify the variables that can quantify the metrics. This would in turn lead to the ability to identify, gather, and analyze relevant community resilience data, culminating in the development of systems-based models to assess and improve community resilience and – via realistic simulations drawn from actual disruptions – support the development of standards, codes, and best practices to enhance resilience.

14.4   Global Initiatives to Support Resilient Communities

On a global scale and largely as a reaction to the various natural disasters in the period from 2004 to 2014, several of which have been mentioned in this book, there has been much attention paid to reducing the risk to populations caused by natural disasters. The Sendai Framework was adopted by UN Member States on March 18, 2015, at the Third UN World Conference on Disaster Risk Reduction in Sendai City, Japan. The Sendai Framework is a 15-year, voluntary, non-binding agreement that recognizes that each nation has the primary role to reduce disaster risk but that responsibility should be shared with local government, the private sector, and other stakeholders. It aims for the following outcome: "The substantial reduction of disaster risk and losses in lives, livelihoods and health and in the economic, physical, social, cultural and environmental assets of persons, businesses, communities and countries." The United Nations Office for Disaster Risk Reduction (UNISDR) is tasked with implementing the Framework, with the following targets (www.unisdr.org/we/coordinate/sendai-framework):

(a) Substantially reduce global disaster mortality by 2030, aiming to lower average per 100,000 global mortality rate in the decade 2020–2030 compared to the period 2005–2015

(b) Substantially reduce the number of affected people globally by 2030, aiming to lower average global figure per 100,000 in the decade 2020–2030 compared to the period 2005–2015

(c) Reduce direct disaster economic loss in relation to global GDP by 2030

(d) Substantially reduce disaster damage to critical infrastructure and disruption of basic services, among them health and educational facilities, including through developing their resilience by 2030

(e) Substantially increase the number of countries with national and local disaster risk reduction strategies by 2020

(f) Substantially enhance international cooperation to developing countries through adequate and sustainable support to complement their national actions for implementation of this Framework by 2030

(g) Substantially increase the availability of and access to multi-hazard early warning systems and disaster risk information and assessments to the people by 2030

Of course, as stated explicitly in the Sendai Framework agreement, the responsibility for the development and implementation of strategies for the reduction of risk from natural disasters rests with each individual nation. If we focus our attention on coastal nations, and more to the point, to major coastal population centers, we discover a very wide range of investment strategies and approaches to the issue. As has already been mentioned, it often appears that whether it is for economic reasons or public policy concerns, communities seem unable to invest in major mitigation projects until an extreme natural disaster has actually occurred. What is required is a forward-looking strategy that addresses not only the vulnerabilities observed during the event, but also the threat(s) associated with other, more impactful events that may have never even occurred at the location before. The key, therefore, is to support a change in thinking at the local scale. And nowhere is this change of thinking more impactful than in the coastal megacities.

In this context, two other global initiatives are worth mentioning here: Rebuild by Design (see www.rebuildbydesign.org/about) and 100 Resilient Cities (see www.100resilientcities.org). Both of these organizations work closely with communities, designers, policy makers, and technology developers from around the world to create networks of knowledge and experience that can lead to more resilient communities.

The 100 Resilient Cities program was created by the Rockefeller Foundation in 2013. The first group of 32 cities was announced in December, 2013 following a rigorous international competition. In December, 2014, another 35 cities were announced as the next cohort. The third and final group of cities was announced in May 2016. Interestingly, and not surprisingly, more than half of the member cities admitted to the 100 Resilient Cities program are located on the coast. The program provides each city with the resources to establish the position of Chief Resilience Officer, who then leads the city's resilience efforts. It also provides access to expert advice and lessons learned from across the 100 member cities and beyond. The organization operates as an independent 501(c)(3) nonprofit organization.

Rebuild by Design is an organization that began as a design competition, launched in 2014 in the aftermath of Hurricane Sandy by the US government in partnership with nonprofit organizations. The program resulted in the funding of resilience-building projects in 13 different cities across the nation, including six projects in the New York metropolitan area, with a total Federal investment of US$1 billion. The success of the program led to the creation of the Rebuild by Design organization, which promotes the use of innovative research and design to assist communities around the world to become more resilient. The organization partners with universities to conduct educational and outreach programs, produces policy papers, and works with the 100 Resilient Cities organization to inform cities around the world about the continuing lessons learned and best practices around the concept of designing for resiliency.

Clearly, efforts to support community resilience require engagement with the appropriate decision-makers in those communities. Usually, this must be accomplished in the context of complex engineered systems in sectors (e.g., transportation) that span multiple political and legal jurisdictions (including international borders) and in which decisions are driven by diverse and sometimes conflicting aims. New resilience designs, tools, and processes will require public and private stakeholders to collaborate. As these global collaborative efforts continue to emerge and expand, there will be a need to coordinate the establishment of metrics and performance expectations. For that collaboration to be effective, institutions and individuals will need to build trust and will also need to establish the goal of community and infrastructure resilience as among the highest priority goals of our generation.

14.5   What Is Next?

The lessons learned, ongoing discussions, and global initiatives outlined here will certainly lead to changes in how cities, and urban coastal communities in particular, adapt to a landscape of natural hazards that appears to be evolving, perhaps rapidly. A "new normal" of more frequent and more extreme weather events, coupled with rising sea levels, all against a backdrop of continued population growth and migration to coastal cities, suggests that the scientific community and the planning and design communities must work more closely than ever before. What is needed is a collaborative effort aimed at developing, communicating, and implementing measures and policies that will produce a built environment along our urban coastlines that is by design and function much more resilient to future natural hazards.

We use the term "function" quite deliberately here. It is our view that in the context of increasingly complex, critical lifeline infrastructure faced with future disruptions that may not have occurred in the past, communities must focus not on the preservation of the **system**, but rather they (and we) should focus on the

function. So, for example, if a marine transportation system is critical to the delivery of food to a particular urban area, it is the delivery of food that should be the focus when anticipating the recovery from an extreme event, not necessarily the repair of the marine transportation system. This focus on food delivery will clearly and immediately identify alternative delivery modes and will point the way toward the development of perhaps new or improved delivery systems, whether they be by rail, road, or alternative waterborne transport. In instances such as this, and in other critical areas such as communication, power, health-care, transportation, and water, we may find that "simpler is better" and the provision of basic but easily recoverable functioning in the aftermath of a disruption due to an extreme natural disaster might be the most effective approach. As we consider this, we should always bear in mind that resiliency is nearly always about people. We must remember that people design the systems that we are discussing here, people operate and maintain them, and they directly benefit from them. No amount of design or preparation can succeed without first understanding how people will respond to extreme disruptions, and further how they will use and/or operate the critical systems on which our modern coastal communities depend.

14.5.1   *Future Academic Programs*

Thus far we have outlined the challenges we face in building more resilient systems and, ultimately, more resilient communities. One question that remains unanswered is this: can a single, perhaps multi-disciplinary academic program be developed that would equip professionals with the knowledge and skills to become leaders in this emerging field?

The University of Tokyo Graduate School of Engineering initiated in 2013 a "Transdisciplinary Education Program on Resilience Engineering". This program includes two required courses: a lecture series on Resilience Engineering and a project. The remaining courses are chosen by the student from among three different tracks: (1) "Fundamental Principles of Resilience Engineering"; (2) "Practical Methods of Resilience Engineering"; and (3) "Social Science for Resilience Engineering". The program is administered by the department of Systems Innovation, but the credits earned under the program are accepted in each of the departments in the Graduate School of Engineering. Students who complete a minimum of 14 credits in the Resilience Engineering program earn a "diploma of Transdisciplinary Education Program on Resilience Engineering". A total of 30 credits are required for a Masters degree in each department.

It is our opinion that the study of Resilience Engineering (or another, perhaps science-based or policy-based degree program) is most appropriately pursued at the graduate rather than the undergraduate level. We believe that any educational program in this domain must be multi-disciplinary, as in the case of the program established at the University of Tokyo. We also believe that the program should ideally include courses and lessons learned from institutions elsewhere in the world.

References

Amante, C. and B. W. Eakins. 2009. ETOPO1 1 Arc-Minute Global Relief Model: Procedures, Data Sources and Analysis. NOAA Technical Memorandum NESDIS NGDC-24. National Geophysical Data Center, NOAA. doi:10.7289/V5C8276M. www.ngdc.noaa.gov/mgg/global/global.html (accessed September 17, 2017).

Amato, D., J. Bishop, C. Glenn, H. Dulai, and C. Smith. 2016. Impact of submarine groundwater discharge on marine water quality and reef biota of Maui. *PLoS One* 11(11): e0165825.

Aon Benfield. 2013. *Hurricane Sandy Event Recap Report*. London: Aon Benfield.

Blake, E. S., T. B. Kimberlain, R. J. Berg, J. P. Cangialosi, and J. L. Beven. 2013. Tropical Cyclone Report: Hurricane Sandy (AL182012), National Hurricane Center, Miami, FL.

Blanton, J. O. 1981. Ocean currents along a nearshore frontal zone on the continental shelf of the southeastern United States. *J Phys Oceanogr* 11(12):1627–1637.

Blumberg, A. F. and G. L. Mellor. 1987. A description of a three-dimensional coastal ocean circulation model. In N. S. Heaps, ed., *Three-Dimensional Coastal Ocean Models*. Washington, DC: American Geophysical Union.

Blumberg, A. F., L. A. Khan, and J. P. St. John. 1999. Three-dimensional hydrodynamic model of New York Harbor Region. *J Hydraul Eng* 125(8):799–816.

Bornstein, R. D. 1968. Observations of the Urban Heat Island Effect in New York City. *J Appl Meteorol* 7:575–582.

Bruno, M. S., R. Boumphrey, F. Lickorish, et al. 2015. Foresight Review of Resilience Engineering, Designing for the Expected and Unexpected. Lloyd's Register Foundation. Report Series: No 2015.2. London.

Clark, R. B. 2001. *Marine Pollution*. Oxford, UK: Oxford University Press.

Cox, W. 2017. The 37 Megacities and Largest Cities: Demographia World Urban Areas: 2017. www.newgeography.com (accessed November 18, 2017).

Cushman-Roisin, B. and J.-M. Beckers. 2011. *Introduction to Geophysical Fluid Dynamics*. Waltham, MA: Academic Press.

DeGroot, W. 1982. Stormwater Detention Facilities. New York, NY: American Society of Civil Engineers.

Dimou, K., T. Su, R. Hires, and R. Miskewitz. 2006. Distribution of polychlorinated biphenyls in the Newark Bay Estuary. *J Hazard Mater* 136(1):103–110.

Doocy, S., A. Daniels, S. Murray, and T. D. Kirsch. 2013. *The Human Impact of Floods: A Historical Review of Events 1980–2009 and Systematic Literature Review*. PLOS Currents Disasters, 1st edn. doi: 10.1371/currents.dis.f4deb457904936b07c09daa98ee8171a.

Ekman, V. W. 1905. On the influence of the earth's rotation on ocean currents. *Arch Math Astron Phys* 2:1–52.

Fischer, H. B. 1972. Mass transport mechanisms in partially stratified estuaries. *J Fluid Mech* 53(4):671–687.

Fischer, H. B., J. List, C. Koh, J. Imberger, and N. Brooks. 1979. *Mixing in Inland and Coastal Waters*. New York: Academic Press.

Flament, P. and L. Armi. 1985. A series of satellite images showing the development of shear instabilities (cover blurb). Eos Trans AGU 66(27):523.

Georgas N., A .F. Blumberg, M. S. Bruno, and D. S. Runnels. 2009. Marine Forecasting for the New York Urban Waters and Harbor Approaches: The Design and Automation of NYHOPS. 3rd International Conference on Experiments/Process/System Modelling/Simulation & Optimization. Vol 1, pp. 345–352.

Gill, S. K. and J. R. Schultz, eds. 2001. Tidal Datums and Their Applications. NOAA Special Publication NOS CO-OPS 1. National Ocean Service, Center for Operational Oceanographic Products and Service, NOAA. https://tidesandcurrents.noaa.gov/publications/tidal_datums_and_their_applications.pdf (accessed November 1, 2017).

Hansen, D. and M. Rattray. 1966. New dimensions in estuary classification. *Limnol Oceanogr* 11:319–325.

Holling, C. S. 1973. Resilience and stability of ecological systems. *Ann Rev Ecol Syst* 4(1):1–23.

Horton, R., C. Little, V. Gornitz, D. Bader, and M. Oppenheimer. 2015. New York City Panel on Climate Change 2015 Report Chapter 2: sea level rise and coastal storms. *Ann NY Acad Sci* 1336(2015):36–44.

Hunter, E. J., R. J. Chant, J. L. Wilkin, and J. Kohut. 2010. High-frequency forcing and subtidal response of the Hudson River plume. *J Geophys Res* 115:C07012. doi:10.1029/2009JC005620.

IPCC (Intergovernmental Panel on Climate Change). 2007. Climate change 2007: The physical science basis. In S. Solomon, D. Qin, M. Manning, et al., eds., Contribution of Working Group I to the Fourth Assessment Report of the Intergovernmental Panel on Climate Change. Cambridge, UK and New York, NY: Cambridge University Press. www.ncdc.noaa.gov/monitoring-references/faq/indicators.php (accessed October 28, 2017).

IPCC (Intergovernmental Panel on Climate Change). 2013. Climate change 2013: the physical science basis. In T. F. Stocker, D. Qin, G.-K. Plattner, et al., eds, *Contribution of Working Group I to the Fifth Assessment Report of the Intergovernmental Panel on Climate Change*. Cambridge, UK and New York, NY: Cambridge University Press.

Kameʻeleihiwa, L. 2016. *Oʻahu is Famous as Land Fat with Food because of Ancestral Teaching that Allows Us to Call Out, "Hello, Come and Eat, and Eat What There Is! Chapter in Food and Power in Hawaiʻi*. Honolulu: University of Hawaiʻi Press.

Keala, G. 2007. *A Manual on Hawaiian Fishpond Restoration and Management*. Honolulu: College of Tropical Agriculture and Human Resources, University of Hawaiʻi at Mānoa. 96822. Available at: www.ctahr.hawaii.edu (accessed June 10, 2017).

Lee, Y.-W. and G. Kim. 2007. Linking groundwater-borne nutrients and dinoflagellate red-tide outbreaks in the Southern Sea of Korea using a Ra tracer. *Estuar Coast Shelf Sci* 71(1–2): 309–317.

Lerczak, J. A. and W. R. Geyer. 2004. Modeling the lateral circulation in straight, stratified estuaries. *J Phys Oceanogr* 34:1410–1428.

LOICZ (Land Ocean Interactions in the Coastal Zone). 2005. *Science Plan and Implementation Strategy*, report prepared under the International Geosphere-Biosphere Programme (IGBP) by Kremer, H. H., M. D. A. Le Tissier, P. R. Burbridge, N. N. Rabalais, J. Parslow, and C. J. Crossland, eds.

Madsen, O. S. and W. D. Grant. 1976. Sediment Transport in the Coastal Environment. Rep. No. 209. Cambridge, MA: Ralph M. Parsons Laboratory, Massachusetts Institute of Technology.

Markowski, P. and Y. Richardson. 2010. *Mesoscale Meteorology in Midlatitudes*. Barcelona, Spain: John Wiley & Sons, Ltd.

Mellor, G. L. and T. Yamada. 1982. Development of a turbulence closure model for geophysical fluid problems. *Rev. Geophys.* 20(4):851–875.

Morison, W.S. 1912. U.S. Hydrographic Chart No. 1262. U.S. 62nd Congress, 2nd session Senate Doc. 575, Panama Canal traffic and tolls.

Nelson, C., M. Donahue, H. Dulaiova, et al. 2015. Fluorescent dissolved organic matter as a multivariate biogeochemical tracer of submarine groundwater discharge in coral reef ecosystems. *Mar Chem* 177(Part 2):232–243.

New Hampshire DES. 2018. New Hampshire Department of Environmental Services website. www.des.nh.gov/organization/divisions/water/wmb/coastal/trash/documents/marine_debris .pdf (accessed June 18, 2018).

Nicholls, R.J., S. Hanson, C. Herweijer, et al. 2008. *Ranking Port Cities with High Exposure and Vulnerability to Climate Extremes: Exposure Estimates*, OECD Environment Working Papers, No. 1. Paris: OECD Publishing.

Nicklow, J. W., P. F. Boulos, and M. K. Muleta. 2006. Comprehensive Urban Hydrologic Modeling Handbook for Engineers and Planners. Pasadena, CA: MWH Soft, Inc.

Nixon, S. W., J. W. Ammerman, L. P. Atkinson, et al. 1996. The fate of nitrogen and phosphorus at the land-sea margin of the North Atlantic Ocean. *Biogeochemistry* 35(1): 141–180.

NOAA (National Oceanic and Atmospheric Administration). 2012. National Ocean Service, Center for Operational Oceanographic Products and Services website. https://tidesandcurrents.noaa .gov/ports/ports.html?id=8518750&mode=threedayswl (accessed October 31, 2012).

NOAA (National Oceanic and Atmospheric Administration). 2014. National Ocean Service, Center for Operational Oceanographic Products and Services website. https://tidesandcurrents.noaa .gov/datum_options.html (accessed October 1, 2014).

NOAA (National Oceanic and Atmospheric Administration). 2015a. Sea Level Trends. National Ocean Service, Center for Operational Oceanographic Products and Services website. www .tidesandcurrents.noaa.gov/sltrends/sltrends.html (accessed November 15, 2015).

NOAA (National Oceanic and Atmospheric Administration). 2015b. March 11, 2011 Japan Earthquake and Tsunami. Updated March, 2015. http://ngdc.noaa.gov/hazard (accessed August 16, 2017).

NOAA (National Oceanic and Atmospheric Administration). 2017a. How frequent are tides? National Ocean Service website. https://oceanservice.noaa.gov/facts/tidefrequency.html (accessed November 1, 2017).

NOAA (National Oceanic and Atmospheric Administration). 2017b. National Ocean Service, Center for Operational Oceanographic Products and Services website. www.tidesandcurrents.noaa.gov/waterlevels.html (accessed January 28, 2017).

NOAA (National Oceanic and Atmospheric Administration). 2017c. US Integrated Ocean Observing System website. https://ioos.noaa.gov/about/regional-associations/ (accessed October 23, 2017).

NOAA (National Oceanic and Atmospheric Administration). 2018a. National Ocean Service, Coastal Services Center website. https://coast.noaa.gov/data/gallery/ (accessed January 15, 2018).

NOAA (National Oceanic and Atmospheric Administration). 2018b. National Ocean Service, Center for Operational Oceanographic Products and Service website. https://tidesandcurrents.noaa.gov/datums.html?id=8651370 (accessed June 10, 2018).

NOAA (National Oceanic and Atmospheric Administration). 2018c. National Ocean Service website. www.noaa.gov/resource-collections/ocean-pollution (accessed June 18, 2018).

NOAA (National Oceanic and Atmospheric Administration). 2018d. National Ocean Service, Center for Operational Oceanographic Products and Service website. https://tidesandcurrents.noaa.gov/faq2.html#10 (accessed June 10, 2018).

Norris, F. H., S. P. Stevens, B. Pfefferbaum, K. F. Wych, and R. L. Pfefferbaum. 2008. Community resilience as a metaphor, theory, set of capacities, and strategy for disaster readiness. *Am J Community Psychol* 41(1–2):127–150.

Ochoa-Muñoz, M. J., C. P. Valenzuela, Si. Toledo , C. A. Bustos, and M. F. Landaeta. 2013. Feeding of larvae of a clinid fish in a southern Chilean estuary. *J Marine Biol Oceanogr* 48(1):45–57.

Oke, T. R., G. Mills, A. Christen, and J. A. Voogt. 2017. *Urban Climates*. Cambridge: Cambridge University Press.

Orlanski, I. 1975. A rational subdivision of scales for atmospheric processes. *Bull Am Meteorol Soc* 56(5):527–530.

Pecchioli, J.A., M. Bruno, R. Chant, et al. 2006. *The New Jersey Toxics Reduction Workplan for New York–New Jersey Harbor: Study I–E – Hydrodynamic Studies in the Newark Bay Complex*. Trenton, NJ: New Jersey Environmental Department.

Pence, A. M., M. S. Bruno, A. F. Blumberg, N. Dimou, and K. L. Rankin. 2005. Hydrodynamics governing contaminant transport in the Newark Bay Complex. In: *Proceedings of the Third International Conference on Remediation of Contaminated Sediments*, edited by S. J. Price and R. F. Olfebbuttel. Columbus, OH: Battelle Press.

Pritchard, D. W. 1954. A study of the salt balance in a coastal plain estuary. *J Mar Res* 13:133–144.
1955. Estuarine circulation patterns. *Proc Am Soc Civil Eng* 81:717/1–717/11.
1967. What is an estuary: physical viewpoint? In: G. H. Lauf, ed., *Estuaries*. AAAS Publ. 83. Washington, DC: American Association for the Advancement of Science, pp. 3–5.

Rasian, Y. and R. Salama. 2015. Development of Nile River Islands between Old Aswan Dam and New Esna Barrages. *Water Science* 29(1):77–92.

Robine, J.-M., S. L. K. Cheung, S. Le Roy, et al. 2008. Death toll exceeded 70,000 in Europe during the summer of 2003. *Comptes Rendus Biologies* 331(2):171–178. doi:10.1016/j.crvi.2007.12.001.

Rockström, J., W. Steffen, K. Noone, et al. 2009. Planetary boundaries: exploring the safe operating space for humanity. *Ecol Soc* 14:32.

Seitz, R. C. 1971. *Temperature and Salinity Distributions in Vertical Sections along the Longitudinal Axis and across the Entrance of the Chesapeake Bay (April 1968 to March 1969).* Graphical Summary Report No. 5, Ref. 71–7. Baltimore: Chesapeake Bay Institute, The Johns Hopkins University, p. 99.

Shields, A. 1936. *Application of Similarity Principles and Turbulence Research to Bed-Load Movement.* Unpublished Ph.D. thesis, California Institute of Technology, Pasadena, CA. http://resolver.caltech.edu/CaltechKHR:HydroLabpub167

Shields, A. F. 1936. *Application of Similarity Principles and Turbulence Research to Bed-Load Movement. Mitteilungen der Preussischen Versuchsanstalt fur Wasserbau und Schiffbau,* Vol 26. Berlin, Germany, pp. 5–24.

Steffen, W., K. Richardson, J. Rockström, et al. 2015. Planetary boundaries: guiding human development on a changing planet. *Science* 347(6223):1259855.

Taleb, N. N. 2010. *The Black Swan: The Impact of the Highly Improbable.* New York, NY: Random House Publishing Group.

Talke, S. A., P. Orton, and D. A. Jay. 2014. Increasing storm tides in New York Harbor. *Geophys Res Lett* 41:1844–2013.

Torab, M. and M. Azab. 2006. *Modern Shoreline Changes along the Nile Delta Coast as an Impact of Construction of the Aswan High Dam.* Philadelphia, PA: World Congress of Soil Science.

Trenberth, K. E. and J. T. Fasullo. 2013. An apparent hiatus in global warming? *Earth's Future* 1 (1):19–32.

UNESCO. 2005. *An Implementation Strategy for the Coastal Module of the Global Ocean Observing System.* GOOS Report No. 148; IOC Information Documents Series No. 1217.

United Nations, Department of Economic and Social Affairs, Population Division. 2015. World Urbanization Prospects: The 2014 Revision (ST/ESA/SER.A/366).

US Army Corps of Engineers. 1948. Cooperative Beach Erosion Study, Atlantic City, New Jersey, December 1, 1948. Philadelphia, PA.

    1984. Shore Protection Manual, Vol. 1. Coastal Engineering Research Center, Waterways Experiment Station, Vicksburg, MO.

    1989. Water Levels and Wave Heights for Coastal Engineering Design. Engineer Manual 1110–2-1414. Washington, DC.

    1998. Coastal Engineering Manual, Part III. 1110–2–292. Washington, DC.

US Environmental Protection Agency. 2004. Report to Congress: Impacts and Control of CSOs and SSOs. Document No. EPA 833-R-04-001. Washington, DC: US Environmental Protection Agency.

USGCRP. 2017. *Climate Science Special Report: Fourth National Climate Assessment,* Volume 1. Edited by D. J. Wuebbles, D. W. Fahey, K. A. Hibbard, D. J. Dokken, B. C. Stewart, and T. K. Maycock. Washington, DC: US Global Change Research Program.

USGS. 2017. US Geological Service website. https://earthquake.usgs.gov/earthquakes/browse/ (accessed November 12, 2017).

USGS. 2018. Water Data for USA. https://waterdata.usgs.gov (accessed May 24, 2018).

Waldman, J. 2013. *Heartbeats in the Muck. The History, Sea Life, and Environment of New York Harbor.* New York, NY: Fordham University Press.

World Meteorological Organization. 2017. www.jcommops.org (accessed September 5, 2017).

Zajic, D., H. J. S. Fernando, R. Calhoun, et al. 2011. Flow and turbulence in an Urban Canyon. *J Appl Meteorol Climatol* 50:203–223.

Zeisel, W. N., Jr. 1990. Shark!!! and other sports fish once abundant in New York Harbor. *Seaport: New York's History Magazine* (Winter/Spring).

Zhang, J. and A. Mandal. 2012. Linkages between submarine groundwater systems and the environment. *Curr Opin Environ Sustainability* 4(2):219–226.

Index